Writing Up

QUALITATIVE RESEARCH

3RD edition

If I lived twenty more years and was able to work,
how I should have to modify the Origin, *and how much*
the views on all points will have to be modified!
Well it is a beginning, and that is something. . .

—Charles Darwin to J. D. Hooker, 1869

Writing Up
QUALITATIVE
RESEARCH
3RD edition

Harry F. Wolcott
University of Oregon

Los Angeles • London • New Delhi • Singapore • Washington DC

For information:

SAGE Publications, Inc.
2455 Teller Road
Thousand Oaks,
  California 91320
E-mail: order@sagepub.com

SAGE Publications Ltd.
1 Oliver's Yard
55 City Road
London EC1Y 1SP
United Kingdom

SAGE Publications India Pvt. Ltd.
B 1/I 1 Mohan Cooperative
  Industrial Area
Mathura Road, New Delhi 110 044
India

SAGE Publications Asia-Pacific
  Pte. Ltd.
33 Pekin Street #02-01
Far East Square
Singapore 048763

Printed in the United States of America

Library of Congress Cataloging-in-Publication Data

Wolcott, Harry F., 1929-
Writing up qualitative research / Harry F. Wolcott. — 3rd ed.
    p. cm.
Includes bibliographical references and index.
ISBN 978-1-4129-7011-2 (pbk.)
    1. Technical writing. 2. Ethnology—Authorship. 3. Educational anthropology—Authorship. I. Title.

T11.W65 2009
808'.0666—dc22   2008028598

This book is printed on acid-free paper.

08   09   10   11   12   10 9 8 7 6 5 4 3 2 1

Acquisitions Editor:	Vicki Knight
Associate Editor:	Sean Connelly
Editorial Assistant:	Lauren Habib
Production Editor:	Astrid Virding
Copy Editor:	Liann Lech
Typesetter:	C&M Digitals (P) Ltd
Proofreader:	Dennis W. Webb
Cover Designer:	Glenn Vogel
Marketing Manager:	Stephanie Adams

Contents

About the Author/
About the Book

Harry F. Wolcott is Professor Emeritus in the Department of Anthropology at the University of Oregon. During his 40+ years at Oregon, he has served on the faculties of education and anthropology, and the intersection of the two fields is where his academic interests lie: Anthropology and Education. After completing his PhD at Stanford University in 1964, he accepted a position at Oregon, and he is still there today. Such long tenure has served him well, for in addition to teaching, he has been able to research and publish a number of fieldwork-based articles and books dealing with topics as varied as the schooling of Kwakiutl Indian children in a small village along the west coast of British Columbia, Canada, and the drinking habits of urban Ndebele and Shona in Bulawayo, Zimbabwe.

The earliest edition of this book, as you will soon learn, came about as a result of a conversation with editor Mitch Allen, who was with SAGE Publications at the time. It appeared as a popular 94-page monograph in the SAGE Series on Qualitative Research Methods. Revised and slightly expanded for a second edition in 2001, it has here been revised once more. Students and instructors alike have commended the book for its user-friendly approach and helpful content, and their enthusiasm prompted SAGE to invite Wolcott to reenergize it again, updating where necessary to keep it current.

The author has been pleased with the book's reception. Recognizing that books about writing can become outdated but seem never to go completely out of fashion, his intent here has been to update as appropriate and to present a more tightly organized manuscript, without anyone having to worry that anything helpful in the earlier versions has been omitted.

Preface

For this third edition of *Writing Up Qualitative Research,* my guiding premise has been "If it ain't broke, don't fix it." Unlike most other aspects of doing and reporting research, guidelines about writing do not become dated except for their citations. Consider, for example, the still popular *Elements of Style* by William Strunk, Jr., and E. B. White, which traces its beginnings to a paper originally written for his students by Professor Strunk in 1918. Matter of fact, if anything I could do or say would help someone to become as successful as Charles Dickens or Jane Austen, I would not hesitate to say or do it, and the world would be better for it. With this edition, I have had another opportunity to revise the manuscript and update references, but I am not embarrassed by the datedness of many of the earlier ones that remain. Each generation has had its share of advice to offer, and there is no need to turn a deaf ear to our forebears simply to give an appearance of keeping up with the times.

One might give thought to the future of publishing—that concern is on everyone's mind today, and for that answer, we might prefer a crystal ball that could render a better picture of what lies ahead for all writers and for books and the whole publishing field. I cannot predict the future, but I imagine that books are here to stay, however modestly or dramatically the form in which they appear may change. I cannot imagine it otherwise, for books have played an important role in my life and I do not see how they could possibly play less of a role in yours—at least for those of you who plan to remain in the academy.

I am aware that writing books is not the same as reading them. Writing is the subject I address here: what someone who has tried a hand at writing and been relatively successful at it can pass on to others who would like to do likewise. I have a bit more to say about this reading/writing distinction, but I will reserve it for an appropriate place in the text.

I did not set out to be a writer; I became a writer rather slowly, because I had to. You may find yourself in the same boat—you realize that you are going to have to do it. If so, I will try to ease the way and offer some advice for you, but I cannot guarantee that we will end up at the same place, for I confess that along the way I discovered I enjoy writing. What luck to find that you like doing something you have to do anyway. I doubt that I will be able to whip up your enthusiasm for writing to that extent, but I may be of some help with suggestions or encouragement for you to get on with your writing or, if you have stalled, to get going again.

Books with a narrow focus like this are sometimes called monographs, a label assigned to publications that deal with a specialized topic. The present work began as a 90-page monograph (still in print), and therefore you will find me referring to it as both book and monograph. If you are preparing a book-length study focused on one group or one particular problem, your writing probably qualifies as a monograph. Academics show a preference for the label monograph when writing about a focused topic; writing a book suggests a work of broader scope. Outside of academia, however, the term is less well known and sometimes gets confused with "monogram" or "monologue." I like to think that I put my unique mark on my writing, but what you see here is a monograph, not a monogram.

By its very nature, however, writing does produce a sort of *monologue*, a monologue in which one individual monopolizes the conversation, albeit in written rather than spoken form. I wince when anyone refers to something I have written as a monologue, wondering if perhaps that is exactly the word they intend. You have no opportunity to tell me why you chose to read this, what kind of information would be most helpful, or where you would like me to begin. I must more or less "make you up" as I go along, trying to anticipate what brings you to this reading and to address those concerns. More important, just as in lecturing, I must try to convince you that I know what I am talking about, so you will attend to the problems that I identify rather than remain singularly preoccupied with your own. In your writing, you must do the same thing: try to gauge, reach, and hold the attention of an audience you are unlikely to ever see. No wonder my colleague Richard Warren once described writing as "an act of arrogance." (I wonder what he thinks about colleagues who presume to tell *others* how to go about it?)

What do you think about my title, *Writing Up Qualitative Research*? Did it reach out and grab you? If you never gave the title a thought, maybe this is the book for you. If the title immediately set you to wondering whether anything can be learned from someone who writes "up" (don't most people write "down"?), or uses a four-word title when three might have sufficed, then you already may be sufficiently word- and style-conscious about many of the things I am going to discuss.

You may find my style casual, occasionally meandering, perhaps at times disconcerting. Most certainly one reviewer did, noting of one of my earlier publications, "For a book advocating, among other things, the art of writing clearly, it is not clear at all. The tone of the book is 'chatty,' and often departs in reminiscing and side tangents." Well, you can't win 'em all. I try to write unpretentiously. In presuming to write about writing, I may have gone overboard in efforts to be informal rather than didactic. What I have tried to do is help you organize and write up *your* accounts by telling you what I have learned from organizing and writing up mine, and to share ideas gleaned from others. Straight talk, writer to writer.

Yet for both your sake and mine, I prefer writing to lecturing, just as you may prefer reading to being lectured to. Here I have time to think through what I want to say and to work—and play—with words until they convey what I intend for them to convey. And here you can put me down without putting me down; if you fall asleep while reading, I need never know.

CHAPTER ONE

On Your Mark ...

Writing is the only thing that ... when I'm doing it, I don't feel that I should be doing something else.

—Gloria Steinem

The original idea for this book came about in 1988 from Mitch Allen. At the time, he was an acquisitions editor for SAGE Publications who was attending the annual meeting of the American Anthropological Association to exhibit SAGE books and to meet with authors. We were introduced by Jean Campbell, one of my graduate students. She had been telling Mitch how she appreciated my writing. Mitch said he was looking for someone to write a monograph for SAGE's new Qualitative Research Methods Series. He wanted something that would address the problems that graduate students and others were having writing up their qualitative studies, which at that time seemed to be flourishing. He asked if I was interested in such a project. I was flattered to have an editor approach *me* with an idea for a book, although writing about writing had never crossed my mind. I accepted his challenge.

But by the time I got back to my hotel room, I had an outline for a little book clearly in mind. My first pass at the writing did not completely satisfy Mitch, but he patiently persisted. I did not recognize at the time the influence he was to have on my writing, but in the next few months, I was able to satisfy both of us with a finished manuscript, and the monograph that resulted became Volume 20 in the SAGE series.

Most of what I had learned about writing prior to that time was the result of doing it, along with some careful—and carefully meted out—suggestions from numerous critic-readers along the way. Three of my own studies had been published, and my writing was becoming known. A few people (including, of course, graduate students like Jean) were telling me that I was a "natural writer." I appreciated being told that I am a good writer (by academic standards, not literary ones), but in my judgment I am not, at least in the sense of doing something that comes easily and naturally.

An honest claim that I can make is that I care about my writing. I work diligently at editing. What others read are always final drafts, not early ones. Pride and perseverance substitute for talent. Although I do not write with a natural ease, I have learned what it takes to produce (final) copy that may make you think I do.

I will say something about introductions later (in Chapter 6), but for now, let's see what it will take to get you to get some words on paper or on your computer screen for purposes of doing a fieldwork-oriented term paper, a thesis, a doctoral dissertation, even an article or book for publication.

From my book title, you know that the focus is on writing what has come to be known as *qualitative* or *descriptive* or *naturalistic* research. This "naturalistic paradigm" goes by various labels. In 1985, researchers Yvonna Lincoln and Egon Guba identified several aliases for the term *naturalistic*—postpositivistic, ethnographic, phenomenological, subjective, case study, qualitative, hermeneutic, humanistic—the list just kept growing. Those of you who pursue any of these strategies or some closely related approach are my intended audience. You constitute a special subset of field-oriented researchers who not only work in a broad qualitative vein (along with biographers, historians, and philosophers) but apply the label "qualitative" or "qualitative/descriptive" to your research, in contrast to those who specify that they are *doing* biography, history, philosophy, and so forth.

WRITING VERSUS OUGHT-TO-BE WRITING

You do think of yourself as a writer, don't you? Or are you an ought-to-be-writing writer, or worse, an ought-to-be-researching researcher who simply can't get started? You are among the latter if you realize that the pressure is on for you to publish but find yourself at a loss even for a research topic. This is sometimes a problem for recent PhDs who accept university teaching assignments in professional schools or applied fields only to find—as they had suspected all along—that advancement depends on sustained "scholarly production" instead. That usually means writing—a professional lifetime of it.

I know there are some ought-to-be-writing types lurking out there. With the growing acceptance of qualitative approaches in such diverse fields as business administration, communications, cultural studies, economic development, education, international aid, leisure studies, nursing, physical education, public health, social justice, and social psychology, academics have been turning to qualitative research out of desperation as well as inspiration. Having reached a stage in their careers when they are expected to publish, these professionals suddenly find themselves inadequately prepared to conduct research. They look for ways to become what they believe they must become: qualified qualitative researchers. Qualitative approaches beckon because they appear natural, straightforward, even "obvious," and thus easy to accomplish. Were it not for the complexity of conceptualizing a qualitative study, conducting the research, analyzing it, and writing it up, perhaps they would be.

This is not a manual on conducting qualitative research or the basics of grammar. There are books aplenty on those topics, standing by ready to help you.[1] Style is part of the writing process; I will add what I can, or, as often, reiterate advice that seasoned authors have offered for years, but I'd like to think you will develop your own style.

My purpose is to help ensure that whatever you have written down in the way of field notes gets written up into a final account, and written so well that your qualitative study is also a quality one. I write not for professional writers but for professionals who must write. And not for those of you who have always delighted in the art of personal expression, but for those of you who write because others expect you to demonstrate what you are prepared to do (if you are in student status) or to contribute to the ongoing research dialogue in your field (if you are completing a doctorate or are a beginning professional).[2]

Although I address a particular subset of researchers, you should find helpful suggestions for any academic writing. But my focus remains on qualitative/descriptive research and on processes related to getting it written up, rather than on related facets (e.g., conceptualizing, research design, conducting the fieldwork, analyzing), despite the fact that these processes are virtually inseparable. No doubt disappointing to some readers, you may find a mechanical cast to much of the advice I offer. Instead of wheedling you to attempt great leaps of intuitive insight, and to write with such panache that you feel that writing is exactly what you should be doing instead of doing something else, you are more likely to find me arguing in favor of pedestrian coding and sorting to construct your study one brick at a time.

Or one bird at a time, to borrow a title from Anne Lamott's *Bird by Bird: Some Instructions on Writing and Life* (1994). It seems that Lamott's brother, age 10 at the time, was overwhelmed ("immobilized" is her word) by the magnitude of a classroom assignment to prepare a report on birds due the following Monday. Father volunteered the comforting advice: "Bird by bird, Buddy. Just take it bird by bird" (p. 19). Good counsel for her brother also provided a ready-made title for her book about writing and the writer's habits. (I'll say more about titles in Chapter 6, including the risk of using a catchy but oblique one like *Bird by Bird,* but I commend Lamott for exquisitely capturing the essence of the message I want to convey.)

Looking Ahead

I assume that anyone able to envision how to proceed from the top down, by introducing and developing an overarching concept, unifying theory, or persistent paradox, will do just that. If you know what you're about, get on with it. If you do *not* know what you're about, then I recommend that you proceed with less flair (and risk) and develop your account from the bottom up. The suggestions I propose—ranging from when and how to begin (in Chapter 2), or how to keep going (Chapters 3 and 4), to what to do by way of tightening (Chapter 5) and what needs to be included in the front and back matter of a book (Chapter 6)—can be regarded not so much as the best way to go about things as ways to get around thinking that you can't possibly do them at all.

I don't follow all these suggestions myself. Some I have tried but no longer use, a few seem like good ideas to pass along, although I have never tried them. Good advice from reviewers and resources consulted for the first two editions has informed this one; the basic ideas have aged well. Some points are raised merely to review the range of opinion or practice extant. I include an immodest number of references to my own writing—writing done over a period of 40+ years—in part to avoid repeating myself but mostly because they are the studies I know best. I wrote them to be read—I take every opportunity to call attention to them.

A brief monograph cannot presume to be the *Compleat Guide to Writing Up Qualitative Research.* For example, extended dialogues have been underway about descriptive research as *text* and the role of critical analysis in the social sciences as part of the postmodern scene, but those issues are not addressed. Neither do I address issues of the *content* of the underlying theory or conceptual basis of your efforts, although I do discuss the *possible* roles these elements can play. Nor have I presumed to anticipate how electronic media will continue to modify the ways we communicate. You probably have as good an idea of what lies ahead, and you certainly have a greater need to know about such things than I do at this stage in life. But it seems certain that ideas will continue to be conveyed through words, and putting words into a form that can be conveyed to others is where the process begins. That remains a virtual reality—in the old sense of the phrase rather than the current one.

My focus is on the *immediate* task: helping you get your thoughts and observations into presentable written form. If you have conducted your research and are experiencing some uncertainty about how to proceed from here, recognize that you cannot possibly have come this far without some idea of what you thought you were up to when you started. I may be able to free you from feeling that you must pay homage to theory or method before you can press on. We'll get to issues of theory and method in Chapter 4.

I am not going to try to convince you that writing is fun. Writing is always challenging and sometimes satisfying; that is as far as I will try to go in singing its praises. You might think of it as comparable to getting up and going to work each day: Some days are more pleasant than others, but regardless of how you feel, you are expected to be "on the job," whether in an inspired state or not. Keep in mind that even if you are writing only one page—or even one paragraph—a day, eventually you will have a working draft in hand. And that's what you will need to get a start on a polished manuscript.

Enough of these warnings, especially to announce what the book is *not* about. I don't know where I picked up the habit of beginning a writing project with so many disclaimers. Probably it started with writing my dissertation. I assumed (incorrectly, as usual) that I was going to have to defend *every word.* A new term crept into my vocabulary: **delimitations**. I don't think I ever used the word again after completing my dissertation, but I'll admit to a great deal of satisfaction at the time in proclaiming all the things that my study was *not* going to deal with.

Inventorying some important delimitations might provide a starting place for your debut into academic writing, especially if you worry that your only writing experience to date has been personal and private rather than subject to scholarly scrutiny. If you aren't ready to declare exactly what your study is about, try listing some of the things it is *not* about. "Delimitations" is a handy category to include in a dissertation and, in some more subtle form, to include in any academic writing. That advice underscores another message that permeates this book: Anything goes that gets you going. During the editing and revising stages, you can decide whether you have overdone the delimitations, but if they have been worrying you, you may as well confront them from the first.

Throughout the text, I stress the importance of revising and editing. Careful editing is the antidote for the lack of giftedness among the huge corpus of us who recognize that we had better write but are not among the better writers. There is little point to writing up qualitative research if we cannot get anyone to read what we have to report, and no point at all to research without reporting.

A writing tip borrowed from Lewis Carroll's *Alice's Adventures in Wonderland:* When you come to the end, stop. I have said enough by way of introduction. Nobody minds short chapters, especially when there is a long one just ahead.

Summing Up: Jump-Start

- Getting something written so that you can begin the necessary editing is a major theme in this book.

- You are unlikely to identify with Gloria Steinem's feeling that writing is the only thing that when you are doing it, you don't feel you should be doing something else. You are more likely to succeed with Anne Lamott's suggested "bird by bird" approach.

- If you can't get started, consider listing what you feel are the important *delimitations* for your proposed study. No limit on this;

make your list as long as need be. Like everything you write, it can always be edited later. Once you've written what your study is *not* about, maybe you can draft a statement of what it *is* about. Later, you may want to share the list with your readers, but for now, keep it to yourself. It is no way to begin a study!

• Anything goes that gets you writing. For now, let's see if there are some ways to get you started. How about making a (personal) list of all the things that seem to be getting in the way of your writing just now. Is the list immutable, or can you confront and "slay" your demons, one at a time?

NOTES

1. For other writer-related problems, see, for example, any recent publication by William Zinsser, who has been writing about writing since 1976, or Diana Hacker, who has been writing about writing since 1979.

2. If you are writing a dissertation, you might also want to see Biklen and Casella (2007); Rudestam and Newton (2007); or Meloy (1993).

CHAPTER TWO

Get Going

Writing comes more easily if you have something to say.

—Sholem Asch

This chapter presumes to help you begin writing. "Get Going," by the way, is a poor title. It offends my ear, and I hope it offends yours. If it does, there's hope for you as editor of your own material, catching phrases during the rewriting process that somehow slip by on your first or second editing.

Some questions that come to mind for organizing this chapter relate to sorting and organizing data: focusing, or deciding where, when, and how to begin. I will touch upon each of these. First, some practical considerations for getting you started.

The moment you generate sentences that *might* appear in your completed account, you have begun your writing. Whether your earliest efforts survive your subsequent pruning and editing is quite another matter and not of major consequence at this point. If what you are writing could *conceivably* make it into the final draft, you can look anyone in the eye and insist that the writing is (finally?) under way. There are all kinds of fancy words you can use to describe what you are doing prior to writing—organizing, conceptualizing, outlining, mulling, or "cranking-up," as Peter Woods describes it (1985:92–97; 1999:15–22). Until your pen or pencil forms sentences on paper, or you achieve the equivalent effect through some miracle of modern technology, you have no basis for claiming either the sympathy or admiration of your family, friends, or colleagues that you are *really* writing.

GET SET

In addition to having something of consequence to write about, the basic requirements for beginning the writing task include setting a time and place for working and, depending on your personal style and the well-formed-ness of your thoughts, either an idea to develop or a tentative plan for proceeding. With a plan, you can start writing for your intended audience. Be discerning about the minimum conditions you require to sustain your efforts. For example, although I still prefer to know there's an adequate supply of Triscuits and cheese on hand when I begin the day's writing, they are not absolute necessities. On the other hand, I cannot write with real—or even threatened—distraction. For me, uninterrupted quiet is essential.

For productive scholarship, I always found working at home more satisfactory than trying to accomplish anything at my university office, which was a place of constant interruption. For some colleagues, home does not provide sanctuary; they use their offices strategically and productively by carefully protecting or scheduling their writing time. Those who must do their writing apart from either home or office seem to survive, so a practical bit of advice is that if you don't have a natural workplace for writing, create or commandeer one. When I must write without interruption while at the university, I gather my materials and head for the library, but I choose a spot in the library, or even a different library (the law library, the science library) where neither the books nor their readers offer much distraction.

Whenever I show visitors around my house, I especially enjoy showing off the room I designed as my study—an attractive, cedar-paneled room featuring a low built-in countertop for my computer and a high built-in countertop that serves as a desk, and with a sliding glass door that opens onto an outside deck. I command a vista of trees and hills through the steady light of north-facing windows, my professional library on crowded bookshelves just steps away. But these are accoutrements of age and resources, not absolute prerequisites. Much of my earlier writing was done with Bic pens on lined yellow pads at a cleared kitchen counter or table, especially during periods of field research conducted far from my accustomed work station. Whenever the going gets tough—the words won't come or don't seem to make good sense—I still revert to those "old ways," writing with a Bic pen on lined yellow paper. Perhaps those comforting reminders help reassure me that I *can* do it, even when the relentless whirring of my computer hums its menacing note of doubt.

Given a choice, I prefer to write where I can spread before me (and leave undisturbed) the materials I want immediately at hand: one or two dictionaries (including an unabridged one), a thesaurus (I prefer looking through a bound copy to the one on my word processing program), and an extra tablet for jotting down thoughts or working through a complex phrase or idea before committing it to the developing manuscript on the screen. In spite of the inconvenience of having to "clear the desk" each mealtime, even if you take all your meals at home, the kitchen table ought to be free for about 22 hours out of every 24, making it one inviting possibility. Plus, the coffee pot and Triscuits are conveniently nearby.

My point is not really about coffee and Triscuits. You must recognize your own writing-related idiosyncrasies and assess their importance. Pamper yourself. What does it take to get you to sit down to write and keep you productively engaged at it, and to ensure that you will turn to it again, preferably at the same time tomorrow? Given realistic options, what is your best workable combination of time and place? Only to the point that coffee, background music, or whatever become distractions need you be overly concerned. I did have a colleague who found writing at home distracting because it was *too* comfortable and convenient, especially with the kitchen nearby. He needed the austerity of his campus office to keep his writing from becoming fattening as well.

Develop as much routine as possible in order to capitalize on the precious moments available for writing. There will never be enough time, or, conversely, everything will take longer than you think. Routine can help you make efficient use of the time you have. Buffer against interruption. Make yourself sufficiently comfortable that you look forward to your writing time rather than dread the thought. Our newspaper carried a story about a local author who writes with his cat on his lap and his dog at his feet. He doesn't like to disturb them, so he credits them with keeping him at the computer for extended periods. If you have enough pets, you might try that, but old-fashioned self-discipline seems preferable.

I attend to chair and desk heights, overhead lights, air circulation, and room temperature. I like adequate space for the ever-expanding body of materials I want close at hand. I finally realized years ago that I tended to bring my work to the kitchen counter in the old house after breakfast because I found the *height* of the counter and kitchen stools to be so accommodating. When I designed the new house, I included a study with a two-level built-in countertop, one a bit higher (39") than the customary kitchen counter, the other conveniently lower (27 1/2") for the computer. Each level has its own chair: a typing chair at the computer and a high drafting stool at the desk. I like to be "up" and sometimes stand rather than sit as I work.

I was amused to discover in Howard Becker's neat little book *Writing for Social Scientists* (1986) that most of us perform some ritual either as a final act of avoidance or as a physical marker for getting started with the day's writing. Some of his students—and some of mine—reported that they showered, sharpened pencils, or vacuumed prior to writing and, as Jan Lewis noted, "There's always the ironing." A new breed who appear able to sit in front of a monitor and go right to work were early to recognize that the inherent playfulness and ever-expanding capabilities of computers offered a whole new set of distractions ("computer fritters," as Jeffrey Nash [1990] referred to them) able to divert the attention of author/researchers with consequences more devastating than the computer viruses that can attack their programs.

Email offers such a compelling source of distraction—especially for anyone who has elected to receive an announcement of each new message as received—that the time-saving features of the computer can be totally frittered away. My initial resolution to that dilemma was serendipitous. The aging computer in my study did not have a built-in modem, so when I decided (years ago and rather reluctantly, it now seems strange to admit) to succumb to email, I received messages on a different computer located in another room. My own computer was connected only to a source of electricity, and my personal warming-up ritual was limited to the time needed to "boot it up." On days when I had a manuscript in progress, I opened immediately to where I left off the day before; there were no messages waiting for me, and there was no way for me to "get out." If it isn't too late, you might consider this arrangement, dedicating one computer solely to manuscript writing. You won't be tempted to see if you've got email waiting if your "writing" computer doesn't deliver it. But I finally upgraded. I now find it almost impossible not to take the email detour before I settle down. I also yield to the temptation to "check my email" anytime I get stuck.

For years, I continued to use the old computer for my back-up files, and with each new virus scare I breathed a sigh of relief: They couldn't reach me there! (Alas, I doubt that anyone would think of using a second computer to back up data these days.)

Go!

I assess the importance each of us assigns to writing by the priority we give it in terms of competing options and responsibilities. My best time for writing is immediately after breakfast, with the promise of several uninterrupted hours. I consider myself to be "really writing" when writing gets that prime time and priority. The important thing is not that writing must occur first in the day, but that it receives first *priority* in scheduling the day's activities. My hat is off to those who arise at 4:30 a.m. to write, or others who do their writing after everyone else in the household is asleep. I am able, and much prefer, to make writing part of my daily routine that takes place during normal hours, rather than making my schedule a constant test of mettle.

A good indicator of commitment is whether you are able to either ignore the telephone's ring or, if others are present to answer on your behalf, you have instructed that you simply are **not available** during the period devoted to writing. I assume that night writers make difficult choices when members of the household review the evening television fare, particularly the specials. Might you record them and reward yourself by watching later?

FOLLOWING A WRITING PLAN

If you need help in organizing your material—or yourself—let me suggest three components that for me comprise a workable writing plan. Two aspects of this plan need to be spelled out explicitly; it is not sufficient to have them floating around as vague ideas you are "mulling" in your head.

A Workable Writing Plan: Your Statement of Purpose

The first component of the plan, we might even label it the First Commandment, is to commit to paper your **statement of purpose**.

You have your purpose well in mind when you can write a critical, clear, and concise sentence that begins, "The purpose of this study (chapter, monograph, article, assignment) is . . ." Although structurally that is a most uninteresting beginning, I know of no better way to help academic writers find, declare, and maintain a focus than to have this sentence up front, not only in their thoughts but in their manuscript. I am impressed by authors who are able to get the message across more eloquently, but the creative rewriting of your statement of purpose can come later. In any writing over which I have been able to exert an influence (as dissertation director, journal editor, or reviewer), I make a strong case not only for putting that flat-footed statement of purpose into the text but also for making it sentence number one of paragraph number one of chapter number one.

A Workable Writing Plan: Your Table of Contents

The second element of the plan is a *detailed* **written outline** or **list of major topics** arranged in the **sequence** in which you intend to introduce them. Schooling ruined formal outlining for me, with misplaced emphasis on indenting, numbering, and rigid rules of unknown origin (e.g., no single item in a subset, there had to be at least two). But the *purposes* accomplished by an outline are what every author needs: a clear distinction between major points and subordinate ones and an orderly progression for presenting them. The point of this step is to develop a sequence for unfolding a story "bird by bird," not simply to get something written down. I realize that there are fieldworkers who claim to carry their studies entirely in their heads, committing no prior outline to paper. After you are properly seasoned, you may want to try it. Not at first.

Developing a proposed **Table of Contents** (subject to constant revision) accomplishes the same purpose as an outline. To me, preparing a Table of Contents also "feels" more like writing than outlining does, particularly for material that may require several major sections. Tables of Contents are also free from the constraints of formal outlining. I am so convinced of their value as a way to organize any major writing project that I asked each of my doctoral students to include a tentative Table of Contents with their dissertation *proposals!* Proposals are ordinarily prepared before the research is begun, so the usual reaction was, "How can I propose a Table of Contents when I haven't even started the research?" Unbowed, I note further, "And I'd like you to estimate the number of pages you intend to devote to each of your proposed chapters."

The lesson to be learned from this exercise is different from what students expect. Their first attempts often reveal more rigidity in their perception of the structure of a dissertation than actually exists. For example, although Chapter Two of every thesis or dissertation ever to be written seems destined to remain the preferred location for a traditional literature review, there is no ironclad rule—even in the otherwise inflexible graduate school at my university—that Chapter Two must be a literature review. Nor is there any rule insisting that an entire chapter be devoted to that topic at all. (I return to this issue in terms of the need to link your study with the work of others, the focus of Chapter 4.)

Another benefit of assigning a seemingly arbitrary number of pages to a nonexistent chapter for a hypothetical table of contents is that students realize they *do* have an *intuitive* sense of the account they intend to develop. Making an estimate of the number of pages to be devoted to each major topic proves not so ethereal an exercise after all. Qualitative/descriptive studies often exceed reasonable expectations for length. Anticipating how to apportion the account among the topics that need to be covered not only helps achieve a sense of the whole but may prevent initial overwriting in introductory sections that argue the "Significance of the Problem" or provide a historical background that eventually must compete for space with the substantive account.

The out-of-pocket expense of copying multiple drafts of a lengthy study gives real meaning to and incentive for an economy of words for students inclined toward lengthy dissertations. Little do they realize that they are unlikely ever again to be completely at liberty to set their own limits as to length. Journal editors often specify maximum lengths and almost invariably return manuscripts (well, return *my* manuscripts) with the comment "Needs to be cut." Book review editors give assignments in terms of a maximum number of words. Publishers typically tell authors how lengthy a book can be, although conventional wisdom would suggest it is the other way around. My instructions for preparing a monograph for the SAGE Qualitative Research Methods Series were to observe "a strict limit of 100–120 double-spaced manuscript pages, which translates to 80–90 printed pages." Given that the second edition was to be "enlarged and expanded," I assumed I could go on and on—until the editor questioned whether I really had *that much more* to add to what I had said originally.

By the time one identifies the topics that *must* be addressed, and assigns a seemingly arbitrary number of pages to each topic, the message is clear and clearly different from what one might have expected. Before the writing has even begun, it is apparent that the space available for the descriptive material will be limited. Given the level of detail ordinarily associated with qualitative study, how can adequate description be incorporated into the study?

For qualitative studies based on observational or interview data, projecting a Table of Contents, complete with estimates of the length of each chapter, leads to one of the most important and paradoxical circumstances of our work. The major problem in writing up descriptively oriented research is not to *get* but to *get rid of* data! With writing comes the always painful task of winnowing material to a manageable length, communicating an essence rather than compiling the bulky catalog that would provide further evidence of one's painstaking thoroughness. The greater one's commitment to letting informants offer their own interpretation of meanings and events— the emic emphasis, as it is referred to in anthropological circles—the greater the proclivity to provide lengthy accounts that dampen the enthusiasm not only of readers but of potential publishers as well. The lengthier a study, the more costly to produce it, and, correspondingly, the greater the risk if it does not attract a wide readership.

A Workable Writing Plan: Your Story and Voice

The third element of the plan need not be written out but does need to be carefully thought out: Determine **the basic story** you are going to tell, who is to do the telling, and the representational style you intend to follow for bringing observer and observed together. An extended discussion of three such "narrative conventions" for reporting qualitative studies—Realist, Confessional, and Impressionist— appeared in John Van Maanen's popular *Tales of the Field: On Writing Ethnography* (1988) and was followed by a decade of lively debate on a topic largely unrecognized prior to that time.

The question of **authorial voice** is critical in qualitative research.[1] When the focus is on the life of one or a few individuals, the problem is compounded when informants are capable of telling their stories themselves, raising doubts about whether we should make our presence known. In quantitatively oriented approaches, and among the more self-consciously "scientific" qualitative types, researchers typically desert their subjects at the last minute, leaving folks and findings to fend for themselves, seemingly untainted by human hands and most certainly untouched by human hearts.

One of the opportunities—and challenges—posed by qualitative approaches is to treat fellow humans as people rather than objects of study, to regard ourselves as humans who conduct research *among* others rather than *on* them. Fieldworkers have usually found it more efficient to assume the role of narrator than to present an entire account through informants' own words (notable exceptions that were groundbreakers in their day include such classics as Leo Simmons' *Sun Chief* [1942] and Oscar Lewis' *Children of Sanchez* [1961]). There is a long-standing preference for having informants render the *narrative* part of the account in their own words, particularly in life history (e.g., Behar 1992; Crapanzano 1980; Shostak 1981; for an article-length example, see HFW 1983a).

Because the researcher's role is ordinarily an integral part of reporting qualitative work, I write my descriptive accounts in the first person. I urge that others do (or in some cases, be *allowed* to do) the same. I recognize that there still are a few academics and academic editors on the loose who insist that scholarly work be reported in the third person. On two earlier occasions, I have had a journal submission edited into impersonal, third-person language without my permission and without the editor even bothering to inform me. I think the practice reflects a belief that impersonal language intensifies an author's stronghold on objective truth.

Science may be better served by substituting "participants" for "we," or "the observer" for "I," but "this writer" has yet to be convinced that that is our calling. A more compelling guideline can be made for matching the formality of the writing with the formality of the approach. Recognizing the critical nature of the observer role and the influence of his or her subjective assessments in qualitative work makes it all the more important to have readers remain aware of that role, that presence. Writing in the first person helps authors achieve those purposes. For reporting qualitative research, it should be the rule rather than the exception.

READERS AND WRITERS

I have come to the (obviously oversimplified) belief that people whose lives are involved with the written word can be divided into roughly two groups: those who mostly read and those who mostly write. Obviously, many literate folk do neither, and a few are remarkable for their accomplishment at both. In the main, however, I believe that people whose occupations require continuing engagement with written words gravitate toward one of two positions. They become preoccupied either with *consuming* words or with *producing* them, not both.

How about you? Do you consider yourself essentially a reader or essentially a writer? Recognize that my dichotomy may be little more than rationalization, for I do not consider myself a reader. That is not to suggest I do not read; rather, in a professional community of readers (scholars, teachers, researchers, students), and speaking relatively, I am neither a voracious reader nor am I "well read." I have always been obliged to read thousands of pages a year—student papers, dissertations, reports, manuscripts, and proposals from publishers, colleagues, funding agencies, and tenure review committees; professional journals and texts (thank goodness for book reviews); and the magazines and books one reads in the effort—or pretense—of keeping up. Except for the daily newspaper, most of my reading is professional, and much of my professional reading is tedious. Only now that I am retired do I find time to read for pleasure.

I read what I must; I write whenever I think I have something important to say. That probably explains why I find field research so appealing: I become actively involved in the process, seeing and hearing and pondering everything firsthand rather than getting it passively and secondhand. I do not envy colleagues whose research forays take them only to the library or keep them glued to a computer screen. Not surprisingly, I regard my most effective reading as the reading I do while I am engaged in fieldwork and/or preparing a manuscript. Writing gives purpose and focus to searching for new sources and reviewing old ones. It provides pegs on which to hang relevant ideas and a basis for deciding what to retain, what to let go.

EARLY WRITING

Hear this: You cannot begin writing early enough. And yes, I really mean it. Would that mean someone might write a first draft before venturing into the field to begin observations or interviews? Absolutely. Read on.

The conventional wisdom is that writing reflects thinking. I am attracted to a stronger position: that writing *is* thinking (see also

L. Richardson [2000] on this point). Stated more cautiously, writing is one form that thinking can take (see also Becker 1986:ix, who in turn cites Flower 1979; Flower and Hayes 1981). Writers who indulge themselves by waiting until their thoughts are "clear" run the risk of never starting to write at all. And that, as Becker explains, is why it is "so important to write a draft rather than to keep on preparing and thinking about what you will write when you start" (Becker 1986:56).

Writing is not only a great way to discover what we are thinking, it is also a way to uncover lacunae in our knowledge or our thinking. Unfortunately, that means we must be prepared to catch ourselves red-handed whenever we seem not to be thinking at all. The fact should not escape us that when the writing is not going well, our still-nebulous thoughts cannot yet be expressed in words.

This is the point where I think "readers" and "writers" part company. Readers compulsively search for more. They are never satisfied that they know enough, and they are hesitant about addressing the writing task until the "knowing" is complete, which it never is. They are intellectually honest. In addition to our awe and respect, they also deserve our understanding, perhaps even our sympathy. Their easily identified counterparts among fieldworkers are those who falter as fieldwork deadlines approach, insisting that they still don't "have enough" to begin the write-up. The familiar rationale has an admirable note of humility: "I'm not *quite* ready."

Do you recall Richard Warren's characterization of writing as an act of arrogance? Can you enter into arrogance and begin writing in spite of the fact that you *know* you do not know as much as you ought to know? Are the words of Clifford Geertz sufficiently encouraging, that it is "not necessary to know everything in order to understand something" (Geertz 1973:20)? If your answer is that you need first to consult six more volumes in the library, or spend six more weeks in the field, before you will be ready, you may possess an enviable capacity for thoroughness, but I have doubts about you as a writer. If you have something to say, can you sit down right now (why not today?) and turn your hand to writing? (No need even to finish this reading just now. However, if you're browsing in the bookstore, you might want to buy a copy to read later. Otherwise, won't you always wonder how it ended?)

An idea I offer to anyone contemplating a qualitative/descriptive study, and especially to those who express concern about how they will write up a study before the research has even begun, is this: Write a preliminary draft of the study. Then begin fieldwork.

You understand that if you follow this advice, the writing you do is only for yourself. But I earnestly believe **you cannot begin writing too early**. Virtually everyone who writes about writing offers similar advice. Hear Milton Lomask's counsel to would-be biographers: "Irrespective of where your research stands, start the writing the minute some of the material begins coming together in your mind. . . . Get the words down. You can always change them" (1987:26, 27).

For a long time, I believed that this idea of writing before even *beginning* fieldwork was original. More recently, I discovered that anthropologist Sol Tax was giving similar advice to his students more than half a century ago. His former student, Edward Bruner, mentions the idea in his own writing (Bruner 1986:147) and subsequently expanded on what he felt to be Tax's intent:

> The spirit in which it was presented was to emphasize that we should say what we know, and then go into the field to test/ check/develop the ideas further or to discard them if they were off the track. . . .
>
> At the time, I did not actually write my dissertation before going to the field, but Sol's suggestion stuck in my mind. . . . It turns your attention to what is new that you are going to discover by fieldwork. So the suggestion was a serious one but I don't think anyone actually did it. [Edward Bruner, personal communication, 1999]

Turning attention to *what you expect to discover* is among the several possible advantages of early writing. Like the "tentative Table of Contents" exercise described earlier, it calls attention to matters of format, sequence, space limitations, and focus. It also establishes a baseline for your inquiry, your own starting point. You will have documented what you believed to be the case, thereby making a matter of record certain biases and assumptions that might otherwise prove conveniently flexible and accommodating were they to remain only as abstractions. Early writing encourages you systematically to inventory what you already know, what you need to know, and how you can go about looking for it.

You also may discover that describing how you believed things were when you began your study offers a good way to begin your written account, especially if "what everybody knows" turns out to be inaccurate or inadequate. We all know, or have preconceived opinions about, far more things than we realize on virtually any topic of professional interest. Writing is a way to access that personal fund of information—and misinformation. Conversing with colleagues is another way, although it is not necessarily as effective. We should not be hesitant to try our ideas with colleagues or encourage them to share ideas with us. But those of us in the word business are smooth talkers—we can worm our way out of a non sequitur almost as effortlessly as we can find ourselves trapped in one. Writing offers a precise and personal way to capture and give concrete form to sometimes conveniently elusive ideas.

Do I follow this practice of "early writing" myself? Yes, in somewhat modified fashion. Except for my initiation into long-term fieldwork—my dissertation study of an Indian village and its school on the west coast of Canada (HFW 1964), from which I returned without the least idea of how to proceed with the writing—I have always turned to writing early. Writing offers a way of tracking what I have understood and calls attention to what I need to find out. However, early writing also presupposes a willingness to let go of words as easily as you generate them, and I find that hard to do. Therefore, my advice for anyone hesitant about writing is to begin immediately, but because I find myself almost too eager to get started, I try to *delay* the moment when I begin writing anything other than field notes. I focus my effort on an immediately prior stage, "tight outlining," getting sources, concepts, examples, sequence, everything lined up and ready to go. I no longer worry whether I can get something on paper; I know I can. Until you enjoy a comparable sense of confidence, consider starting earlier.

Here is how early writing worked for one of my more skeptical doctoral students, Ben Hill:

> Beside the tentative table of contents, which I had written at Harry Wolcott's suggestion, the greatest impetus for my ethnography-as-writing orientation was Harry's subversive suggestion that I complete a first draft of my dissertation <u>before</u> beginning fieldwork.
>
> The very idea—not that I seriously considered executing it—that an ethnographer might write before doing fieldwork dramatized to me that I was not starting with a blank slate. I had definite preferences on the contents, organization, and style of ethnographies in general and mine in particular. I had expectations, some

warranted and some not, about what I would encounter and learn in Japan. I had biases which inclined me toward interviewing certain types of informants and drawing certain types of conclusions. In some sense I already possessed a draft of the study, an unwritten "zero draft," from which my completed study would emerge through a process more akin to textual editing than to pristine search and discovery. [Hill 1993:102–103]

Freewriting

There are different approaches to ensure that you "get going" with writing. The recommended strategy for anyone whose style has not yet evolved is simply to let the words flow: make no corrections, check no spelling or references, don't even reread when you are on a roll (see Becker 1986, Chapter 3, for elaboration on this technique, known as **"freewriting"**—forcing yourself to "write without stopping for ten minutes"; see also Elbow 1981:13–19). With the miracle of word processing so conveniently at hand and so forgiving about mistakes and changes, freewriting has become possible for everyone. Whether you prefer to "talk" to (freewrite for) only yourself, to save systematically all the bits and pieces for later review, or to pursue every idea to final prose form, your computer awaits your command.

Of course, word processing has created some problems of its own, one of which has become evident in reading student papers (and sometimes colleagues' papers as well). The ease of production can result in faster rather than better writing. Computer capabilities for easy revision and checking for spelling errors are often ignored. Hastily written and hastily proofed first drafts are tendered as final copy; printout is equated with "in print," the sketch proffered in lieu of a more careful rendering. Writing done in student status is unquestionably the worst circumstance for learning to write well. Student writing is usually done on a hurried, one-shot basis, with neither time nor motivation for the reflection and revision that lead to better drafts. The entire process is short-circuited. We want our students to become accomplished writers but seldom provide opportunity for them to develop or practice better writing habits.

As a teacher, one way I found for improving the quality of written work was to offer an "early bird" option for students in classes in which a term paper or research report was required—and a good deal of writing was required in all of them. (I stopped giving exams years ago; I am interested in what students can do, not what they can memorize.) Students able and willing to meet an early deadline had the option of submitting a working draft of their term paper for appraisal well before the final assignment was due. My editorial suggestions may have been of some help, but the real advantage was the preparation of an early draft while time still remained for reflection and revision.

The Methodical Approach

The opposite of freewriting is observed by writers who are sometimes referred to as "**bleeders**." I do not know the origin of the term, although it brings to mind an observation attributed to sports journalist Red Smith, "There's nothing to writing. All you do is sit down at a typewriter and open a vein."

Bleeders are methodical. Their approach reflects a combination of confidence and command about writing, along with some personal qualities (hang-ups?) about having everything just so. They worry over each sentence as they write. They do not press ahead to the next sentence until the present one is perfected. Once properly in place, each sentence is viewed pretty much as a finished product. In the old days, bleeders usually wrote with pencil or pen to facilitate cross-outs and "interlining," squeezing corrections between existing lines. Bleeders tend to be slow writers, but they get the job done. Often, they set a number of words or pages as their daily objective, such as Peter Woods' "standard 'production rate' of five written pages, or a thousand words a day" (1985:93).

Most of us fare better by committing ourselves to blocks of time rather than to a predetermined number of words or pages. In writing up qualitative research, page production by itself can prove a deceptive goal. One might draft ten pages of descriptive narrative one day and struggle with ten sentences of interpretation the next. Nevertheless, if you recognize the bleeder tendency in yourself and cannot imagine romping through an early draft, subsequently discarding material with abandon, then perhaps a tightly detailed outline (or Table of Contents) is sufficient to get you started on the slow-but-steady production of a first draft.

Most likely you will shift back and forth between these approaches. Your progress may depend on mood and energy level, but is more likely to reflect the type of material you are writing and your previous experience. My writing sometimes flows easily, yet it can slow to a snail's pace whenever I encounter trouble at making good sense or am wrestling with an interpretation. When the words don't come easily, I find myself turning from the computer keyboard and retreating to my Bic pens and yellow pads to squeeze words onto paper one at a time.

In *Writing for Social Scientists,* Howard Becker has a chapter seductively titled "One Right Way" (1986: Ch. 3). His point, as experienced writers will recognize, is that there is no such thing! For a comparable chapter in *Writing With Power,* author Peter Elbow takes no risk that anyone might misunderstand; his title is "The Dangerous Method: Trying to Write It Right the First Time" (1981: Ch. 6). Whatever combination of steps and strategies works for you is "right" as long as your ideas are put to paper. Given time, they can be transformed into a more coherent and polished statement. Restrict your hasty writing for email—but hey! even email messages deserve to be proofread.

WHERE TO START THE ACCOUNT

If you have a good sense of how your writing project is to proceed, or you actually prepare a rough draft of a study before beginning your systematic research, then writing is already an integral part of your research agenda; you are in the so-called catbird seat. But don't expect the parts to come together that easily, the writing simply to "flow." If such results could be achieved effortlessly, there wouldn't be so many how-to books and courses about writing, or audiences anxious to have the secrets and recipes of successful writing revealed.

Suppose you are an ultra-conservative researcher who takes seriously the challenge for exploration and discovery inherent in qualitative research. You begin with a broadly defined purpose, acknowledging that you are not sure exactly what you are looking for. To convince yourself of your objectivity, you steadfastly refuse even to acknowledge your hunches, your suspicions, or, to borrow anthropologist Branislaw Malinowski's oft-repeated phrase, your "foreshadowed problems" (Malinowski 1922:9). Only when the fieldwork seems virtually complete do you feel it appropriate to turn attention to writing. (Too bad to have garnered this impression from Malinowski. He actually suggested that the writing should begin earlier, through a "constant interplay of constructive attempts and empirical checking" [p. 13].)

Returning to advice from *Alice's Adventures in Wonderland,* I suggest that you begin at the beginning, continue till you reach the end, then stop. That may be sound advice for how the completed work should read, but it won't prove helpful if you haven't figured out just where your account does begin. Let me suggest possible places for starting.

POSSIBLE PLACES TO BEGIN

One possible place to begin is at the point when you first entered the scene and then to report chronologically from that moment. Another is to begin with description but to underplay your own presence or involvement. Yet another alternative is to begin by writing how things turned out. However, if you can start there, either you are an old hand at this or you have confused qualitative research with gathering support for a position paper. If you are pursuing social reform in an activist mode, in an effort to make research useful and relevant, then you must examine and reveal the sources of your passion with the same scrutiny.

Or you can start with a personal narrative through which you introduce the study in the manner that you actually experienced it, reaching as far back as you feel necessary to put things in context. Your description can begin with how you happened to become interested in the underlying issue or how you approached the setting initially. You know how you went about the study, and your readers might be interested to know, too. Drawing readers into the account the way you were drawn into the setting offers a natural way to unfold the story, with a ready-made sequence to follow.

Starting With Method

You may find writing about fieldwork so inviting that you are tempted to go on and on about it. No harm done if you overdo it a bit at first, especially if the writing helps you find your "way in" to the substance of your study. However, as I discuss in Chapter 4, I recommend that you not devote undue attention in the *final* version to discussing "methods." If you feel the urge for an extended discussion, either about method in general or about how you conducted your research or analyzed the data for a particular study, consider presenting that material in a separate account. There is no longer the need to defend qualitative research or to offer the detailed explication of its "methods" that we once felt obligated to supply.

Such was not always the case. In 1966, I began fieldwork that culminated in *The Man in the Principal's Office: An Ethnography* (HFW 1973). One year later, while continuing fieldwork on a more limited basis, I began writing. Not only did I begin by writing a (cleverly titled) method chapter, "A Principal Investigator in Search of a Principal," I also made it the opening chapter in the completed monograph. In those days, I felt that I first had to explain—and in a sense, to defend—the ethnographic approach I had taken. In a separate article based on that fieldwork, I did the same thing again, carefully explaining my fieldwork procedures before introducing any descriptive material (HFW 1974b).

Today, your discussion of method might be relegated to an appendix in a monograph. For a chapter-length article, a single paragraph may be adequate. My hunch is that if you go on and on about method, whoever is looking over your shoulder (editor, dissertation advisor) is likely to want less rather than more. (If you are in doubt about such expectations, you could always ask!)

Your readers do not need a long treatise on how studies like yours are *usually* conducted. What they need to be informed about is the nature and extent of your *particular* data base. During exactly what period of time did you conduct your research? Assuming you did fieldwork, how extensive was your involvement? In the best of fieldwork traditions, did you reside at the site, or did you commute to it, as I have found myself doing in recent years? To what extent do interviews constitute part of your data base, and, for your purposes, what constituted an interview?

You may also want to say something about "**triangulating**" your data. The practice of checking multiple sources is often touted as one of the strengths of fieldwork. Triangulation is one of those ideas that sounds great in a research seminar but can pose problems in the field. Wait until your informants find out that you are double-checking everything they tell you! So, how did you go about confirming information without simultaneously antagonizing your informants?

Where and how to include such information is partly a matter of personal preference. The important thing is to be up front about it, but, as I argue in Chapter 6, that does not necessarily mean putting it up front. Regardless of where you review your research strategy, I think it judicious to examine and, as appropriate, to qualify any and every statement a reader might perceive as a generalization that does not have a corresponding basis in fact. The phrase may get overworked, but scholarship does not suffer when a sentence begins with "As one villager commented . . ." rather than with "Villagers said. . . ."

As a somewhat idealistic guiding principle, consider Taylor and Bogdan's restatement of this perennial concern: "You should give readers enough information about how the research was conducted to enable them to *discount* your findings" (Taylor and Bogdan 1984:150; see also discussions on establishing *trustworthiness* in qualitative inquiry in Lincoln and Guba 1985:289–331; Denzin and Lincoln 2000:158). Albert Einstein was properly cautious in his purported observation that "no amount of evidence can prove me right, and *any* amount of evidence can prove me wrong" (noted in Miles and Huberman 1984:242).

Starting With Description

Describing how you went about your research may be a good way to get *you* started, and it can be comforting to have a section completed, if only in draft. Unless you are going to develop the account chronologically, however, your reader is more likely to want to get right to the heart of your study. So although you may choose to *write* first about method, the account itself should probably begin with description. What is the problem you address? What is the setting or circumstance in which you addressed it?

Description provides the foundation upon which qualitative inquiry rests. Unless you prove to be gifted at conceptualizing or theorizing, the descriptive account will usually constitute the major contribution you have to make. The more solid the descriptive basis, the more likely it will survive changing fads and fashions in reporting or changing emphases in how we derive meaning from our studies. Give your account a firm footing in description.

If you are comfortable in the role of **storyteller** (and you *do* have a story to tell, if you can bear to regard the reporting of research that way), here is an opportunity to assume that role, inviting the reader to look—through your eyes—at what you have seen. Start with a straightforward description of the setting and events. No footnotes or academic asides, no intrusive analysis, just the facts, carefully presented and interestingly related at an appropriate level of detail.

The vexing question as to just what the appropriate **level of detail** is in a descriptive narrative has no pat answers. Your purposes in conducting the inquiry are your best guide, although they may need to be tempered with attention to the *art* of storytelling. The reactions of invited reviewers may prove especially helpful once you begin revising your draft. Having been immersed in the research setting, you may be unaware that you have omitted details that have become commonplace to you but are not apparent to readers unfamiliar with the setting. It is easy to lose track of abbreviations, acronyms, and assumptions that prevail in professional dialogues and regional dialects.

Your description can begin either with the setting, giving an account of a specific event, or through introducing one or more key players, perhaps letting someone else tell a personal story if readers are not going to be privy to your own. You may be tempted to "wax poetic" in your opening statement, but it is always a letdown to discover that two paragraphs into an account, the tone changes. If you have only two paragraphs of flowery discourse in you, save them for another day and simply get on with some straight descriptive reporting.

An equally vexing problem, one that catches many an unwary observer by surprise, is the subtle but critical distinction between **observed** and **inferred** behavior. Here is an instance where interview data present less of a problem than reporting what we observe with our eyes. What people say can be relayed exactly as they said it. That does not necessarily make it true, but the words themselves can be transcribed and reported as stated (without the subtleties of inflection, body language, etc.). By contrast, what we *see* tends to be interpreted even as we see it. Although we mean to describe observed behavior, we all-too-easily slip into reporting inferred behavior, with action and intent colored by the eye of the beholder.

Only from your own perspective can you report how anyone "felt" about what was happening or the "meanings" they attributed. Unless others specifically express such feelings, what we report should deal with what we actually have seen and heard, never with what we infer unless we are careful to qualify our observations. There is a world of difference between reporting that there *was* a sense of excitement and apprehension in the air or that one *sensed* an atmosphere of excitement and apprehension. Careful description calls for a sense of detachment. If you can't achieve that detached state, or do not want to present your account from so dispassionate a perspective, then you will have to frame your observations in the first person: "Here is what I saw, presented in terms of what I made of it."

Description, in the sense of attaining "pure" description—sometimes lightheartedly called the doctrine of *immaculate perception* (Beer 1973:49)—isn't such straightforward business after all! Without realizing it, even as we describe we are engaging in analysis and interpretation. So the suggestion that you "stay descriptive" as long as possible presents a challenge, not so much to try to achieve "pure" description as to resist the urge to begin analyzing until you have presented sufficient data to support that analysis.

CONTINUING INTO ANALYSIS AND INTERPRETATION

In *Transforming Qualitative Data* (HFW 1994), I proposed a distinction between *analysis* and *interpretation* that I repeat here. This distinction gives analysis a more limited, more precise, and more clearly defined role than is suggested by its broader use as a cover term for anything we do with data. Contrasted with the somewhat freewheeling activity of interpretation, I restrict analysis to refer to the examination of data using systematic and standardized measures and procedures.

In spite of the philosophical musings of the postmodernists, there is a "there" out there. **Analysis**, used in this narrower sense, follows standard procedures for observing, measuring, and communicating with others about the nature of what is "there," the reality of the everyday world as we experience it. Data subjected to analysis are examined and reported through procedures generally understood and accepted in that everyday world, among social as well as not-so-social scientists. Virtually all data amenable to statistical treatment or that can be plugged into a software package fall under the rubric of analysis in this definition of the term. The reliability (in any sense of the word) of such procedures derives from the standardization of procedures, not their rightness or wrongness, nor even their appropriateness.

Content analysis serves as a good example of analysis in this more restricted meaning of the term. In content analysis, material can be chunked into categories and reported statistically through procedures generally understood and accepted, in spite of whatever discrepancies occur in coding. Even seemingly straightforward procedures like averages or percent figures can be manipulated and abused, but it is not because of disagreement about the ways that average or percent figures are derived.[2]

Interpretation, by contrast, is not derived from rigorous, agreed-upon, carefully specified procedures, but from our efforts at sense-making, a human activity that includes intuition, past experience, emotion—personal attributes of human researchers that can be argued endlessly but neither proved nor disproved to the satisfaction of all. Interpretation invites the reflection, the pondering, of data in terms of what people make of them. The basis of symbols and meanings upon which anthropologists derive patterns of cultural behavior, for example, can be described and examined analytically, but discerning the patterns themselves is a matter of interpretation.

Analysis falls more on the scientific side of things, interpretation on the humanistic side. Your leanings toward one dimension or the other should be evident from the links you make with the literatures extant, your "quoting circle," how you "chunk" your data, the relative emphasis you give to measuring and to measures. The processes are not antithetical. A well-balanced study can show ample evidence of attention to both the methical results of analysis and the conjectural tasks of interpretation. But "well-balanced" does not mean trying to achieve a perfect balance between analysis and interpretation. The problem itself probably calls for attending more to one or the other. A clear statement of purpose(s) is critical for deciding what data need to be reported, what needs to be counted, what relevant literatures to cite or measures to use, and how broadly to draw implications or recommendations from one's research.

We have outgrown the guiding maxim of the early positivists that if it can't be measured, measure it anyway! But if you are relatively new to this work (and you probably are, if you are reading this), you are better advised to err by overdoing the analytic dimension rather than to assume that qualitative research automatically bestows poetic license! By exhibiting some analytic prowess, you validate your credentials as a systematic observer who has recorded and examined data with care and mustered the requisite evidence before proceeding ever so tentatively to offer your interpretation. Count and measure whatever warrants being counted and measured. Far better to offer too much measurement data than too little, as long as you aren't using data simply for effect (see Chapter 5).

Don't be surprised if, despite a concerted effort to keep them separate, description and analysis tend to meld as the account unfolds. It is only through the examination of data that data themselves take on meaning. To make sense, you have to start combining things, aggregating data, discerning patterns. But you can treat your initial analytic efforts lightly at first, in service essentially to description. Don't be in a hurry to move beyond the descriptive task in order to get on with what may seem a higher order of business.

Once the descriptive account is firmly in place, I suggest that you proceed with analysis in a manner that keeps it distinguishable from the descriptive material on which it depends, easily identifiable as something you are doing to the data rather than something inherent in the data themselves. Use separate paragraphs or frequent margin headings as necessary to mark shifts in the presentation, especially if you are adding additional descriptive detail in support of the analysis. Otherwise, readers may feel they are being bounced back and forth like ping-pong balls when each new element of description is subjected immediately to heavy-handed analysis or linked with what some previous researcher has found or argued. Once you turn to analysis, any additional descriptive material ought to be immediately relevant to the account you are developing.

When you do turn to analysis, make that dimension of your study as strong and as systematic as possible, the justification for your effort. Be factual in what you report; save the controversial or contestable for your interpretative comments. You might think of the analysis you offer as earning merit that you can later redeem for the opportunity to offer your interpretations.

Sometimes, members of a dissertation committee may flat out advise that they are far less interested in what *you* make of things than with the original data you present and your tentative, close-to-the-vest analysis of those data. They will entertain a certain amount of speculation only after you have demonstrated that your study rests on a solid foundation of description and hard-nosed analysis. Ironically, should you learn that initial lesson too well, you may be surprised in subsequent writing to have an editor press for just the opposite, pushing for broad interpretive remarks when you assumed you were supposed to stick closely to your data. Thus, your "station" enters into what is expected of you or how much freedom you have to interpret your data. What you might not be allowed to do as a beginning researcher may be expected of you once you are recognized as an experienced one.

Until the analysis is well under way, it may be difficult to know how much to include in the descriptive narrative. For starters, I think a good case can be made for drafting the descriptive part of the account prior to beginning the analysis, prior even to determining what the course of that analysis will be. (Extraneous materials can always be cut, because descriptions usually need editing for brevity.) Descriptive material written prior to intensive analysis can provide a check against the analysis itself: If the facts don't fit, something must be wrong with the interpretation. We may end up with *unexplained* findings in qualitative work, but we need not fear the *unwanted* ones that sometimes plague our more quantitatively oriented colleagues. Good qualitative research ought to confound issues, revealing them in their complexity rather than reducing them to simple explanation. If I have learned one thing from the experience of qualitative research—and the experience of life itself—it is that human behavior is *overdetermined.* Our studies should underscore that observation. Let researchers of other persuasions pursue single-issue answers to complex questions.

Whether and when to meld **interpretations** into the account as you develop it or to try to keep analysis and interpretation separate is, once again, a matter of storyteller strategy, personal style, prior experience, and "station." Whatever your decision, do not pretend to be above the fact that there is no such thing as *pure* description. Distinguishing among description, analysis, and interpretation is a matter of emphasis. Were we not selective, and thus subjective, in our focus, we would not be able to construct our accounts at all. Without some preconceived idea of what is to be described, there can be no description. Every step of the way—from setting a problem and selecting an appropriate place, person, or group for studying it, to selective focusing within that setting, to decisions about what gets recorded and which elements of the recorded material find their way into the final account, to the style and authorial voice for accomplishing your purposes—reflects both conscious and unconscious processes of focus and selection.

The less theoretically inclined among us stake our reputations on solid ("thick," whatever that is) description, but we all have been socialized into the subtle norms of the various disciplines that guide our resolution to the question of how much description is enough.

My assessment of qualitative studies in the field of education is that they reveal a tendency toward heavy-handed, intrusive analysis because that is what researchers in that field feel their readers expect. Schooling is so common an experience that readers are assumed to be impatient with lengthy accounts of "what everybody knows." They do not expect or want to be told again. It is a testimonial to good reporting when a researcher presents so compelling an account that the descriptive material draws attention. The problem is further compounded by educational researchers who feel they not only know their educator audiences but know what is best for them. Informants in such accounts do little talking, the researcher does a lot. Each reported observation or quotation seems to prompt comment or interpretation on the part of the omniscient researcher, something like the chatty docent or guide who *becomes* rather than *leads* the tour and has assumed that without such a monologue, we would not know what to think.

I dub studies that exhibit intrusive analysis "Grounded Theory— But Just Barely." In place of the careful grounding in observed behavior that we expect in qualitative inquiry, such studies seem only to skim the surface in their rush to explain how things mean. In a slight variation on this approach, researchers draw back the curtain to let us watch events unfold, but constantly interrupt with scholarly interjections, as if duty bound to remind us of their presence and superior vision.

To be able to meld description and interpretation is a worthy achievement, but at the least suspicion that your analytical or interpretive asides are interrupting rather than advancing the narrative, I suggest you make a renewed effort to keep them apart. Initially, you might set your interpretive comments in parentheses (or *italics* or **boldface**). If you are writing with a word-processing program, you can determine at a later time whether to leave them where they are, relegate them to footnotes, or collect them under a new heading where you mark a shift from descriptive to analytical mode.

If you have not become aware of how—and how easily— description segues into analysis, pay attention to the ways other researchers handle the interplay between observational or archival data and their academic tradition.[3]

You may be surprised, and perhaps disappointed, to discover that some of the studies you previously regarded as exemplars of descriptive work have built upon a conceptual framework apparently well in place before the research began, with case study data playing only an illustrative role. (Small wonder, then, that data and analysis seem to blend so effortlessly.) That is a different, highly selective way to use qualitative data, something of a complement to the descriptively based approach I am discussing—and advocating— here. With conceptually oriented studies, a too-leisurely meander through descriptive material can be distracting, just as intrusive analysis can be distracting to a reader expecting a descriptive account. Remain clear in your own mind what you have set out to do, and make sure the reader clearly understands your intent.

A Reminder About Purposes and Delimitations

If you begin by devoting attention to the descriptive account, let me note as exceptions two short statements that can and should be drafted early, their location in the completed manuscript to be determined later. One is the previously discussed statement of purpose, a candidate for the opening sentence for all scholarly writing, "The purpose of this study is. . . ." The other is a broad disclaimer in which you acknowledge the limitations (or "delimitations," as mentioned earlier) of your study: that it occurred in a particular place, at a particular time, under particular circumstances; that certain factors render the study atypical; that limited generalization is warranted, et cetera, et cetera.

Such a litany of limitations, generally applicable in all qualitative research, might bear repetition in reporting any element that could be misconstrued as unwarranted generalization. The idea behind making all of this explicit, and in doing it early, is that, having said it *once,* you do not have to repeat it for every new topic you introduce. You might think of it as a form of academic throat-clearing. Having stated your disclaimers emphatically, you will find it a great relief not to have to begin each sentence with, "Although this is a case study, and limited generalization is warranted. . . ."

THE PROBLEM OF FOCUS

I keep returning to the importance of the critical sentence, "The purpose of this study is. . . ." What if you can't complete that sentence, because that happens to be the point at which you are stuck?

If that is where you are stuck, writing is not your problem. Your problem is conceptual, one that George Spindler labeled "the *problem* problem." If you don't feel you can make adequate headway with your "*problem* problem" by simply staring into space, you might try either of two approaches.

One approach is to invite a colleague with a good analytical turn of mind to have lunch with you. Order lightly so you can dominate the conversation, chatting about research concerns. Solicit help for your problem of focus. Your colleague may not prove as helpful as you hoped, but giving words to previously unexpressed thoughts may help *you*. When other people, with other thoughts on their minds, offer feedback, even if what they say widely misses the mark, you may discover that you are closer than you realized to pinpointing your ideas. Do consider drawing upon a wider network of colleagues than those who may first come to mind. Graduate students can be a great sounding board, not only for fellow students but for professors. Similarly, professors can be helpful to students, including students who are not their advisees. I'm serious about lunch, to get you away from interruption and from locales where status is fixed. What I said to students in my office always sounded a bit stilted; over lunch, my ideas had to fend for themselves (as well as compete with the french fries).

The other alternative (you might consider both) is to reinterpret the writing task as a way to *resolve* your "problem problem" rather than as the source of it. With only yourself in mind as audience, try the freewriting described earlier. Think on paper. Pin down your thoughts by giving them what Becker calls "physical embodiment" (Becker 1986:56). You may discover that a scholarly explication of the *problem* is necessary prior to proposing new empirical research. Possibly, you can develop a stimulating article (or organize a good symposium) because the problem itself has never been well formulated.

It may be helpful to think of research as problem "setting" rather than problem "solving." In much applied research, the outcomes do not set well because the problems addressed are not well set. In this regard, we old-timers miss opportunities to model good research practice for our students. Renewed enthusiasm for a qualitative approach, especially as it has come to invite more open-endedness in what we look for and how we proceed, has exacerbated the problem of problem setting (also referred to as problem finding or problem posing—see HFW 1988). It makes no sense to go off to "do research" without an idea of what is to be researched, even if the research is subsequently redirected or refocused. Empty-headedness is not the same as open-mindedness.

My bias may be showing, but it has always seemed to me that ethnographers have it better than most other qualitative researchers in this regard because they have such a broadly stated mandate to guide them. They study culture in general, particular aspects of it in particular: cultural themes, culture change and adaptation, political economy, social structure, worldview. I have described this broad charter as "ethnographic intent," arguing that purpose rather than method lies at the heart of all research (HFW 1987, 1990a, 1999a, 2008a). In some quarters, qualitative research—even ethnography itself—has become synonymous with "going to have a look around." No wonder inexperienced researchers have trouble writing up their studies when they set out with objectives so loosely defined or settings mistaken for problems.

I do not mean to create an impression that a research focus is something sacred, that once declared it warrants unwavering commitment, or that, once fixed, the course of a research project cannot be altered. Part of the strategy of qualitative inquiry—a key advantage of the flexibility we claim for it—is that our research *questions* undergo continual scrutiny. Nothing should prevent a research question or problem statement from going through a metamorphosis similar to what researchers themselves experience during the course of a study. Data gathering and data analysis inform the problem statement, just as the problem statement informs data gathering. A research proposal is only that: a *proposal,* a beginning, a starting place, literally a point of departure suggesting how one might proceed. We are burdened with strictures that we ourselves have built into our expectations about the proper conditions of research. A statement attributed to biologist Paul Weiss helps me maintain perspective in these matters: "Nobody who followed the scientific method ever discovered anything interesting" (quoted in Keesing and Keesing 1971:10).

PROBLEMS OF SORTING AND ORGANIZING DATA

If your data remain in essentially the same form in which you originally collected them, pages and pages of notes and interview protocols, I hope you don't attribute your problem to "writer's block."

Your blockage has occurred at an earlier stage. If you have embarked on a descriptive broadside, you had better get back to some very basic sorting into some very basic categories and then see if you can discern some very basic questions to guide the development of your account.

Some questions that guide me (but may not work for you) are: What is going on here? How do things happen as they do? What do people in this setting have to know (individually, collectively) in order to do what they are doing? And, in the absence of explicit instruction, how are necessary skills and requisite attitudes transmitted and acquired?

Such questions guide my thinking because they address processes of culture acquisition, the underlying concern in my academic work. I do not have a problem with focus. Any problem I have is with people who do not share my sense of excitement that these are wonderful questions or who have no equally wonderful (to them, at least) questions of their own to examine. What are the questions that guide you, either of the intermediate sort that guide your current academic pursuits, or the overarching type that shape an entire career?

When you are ready to begin some initial data **sorting**, start by identifying a few of the broadest categories imaginable. How about place and/or time and/or actors as a start? You might begin your study with a snapshot in which you set the scene and introduce major actors one by one, much as if you were writing a play. Continue presenting "still shots" until you have enough elements on hand (or actors figuratively on stage) to put things in motion. This is how George Foster went about examining processes in applied anthropology in a book by that title written years ago (Foster 1969). Foster looked first at what he called the target group, the people targeted for change. Next, he did a predictable anthropological turnabout and directed attention to the innovating organization or donor group, the so-called agents of change. What was it these change agents hoped to achieve, both for the target group and for themselves? Having examined each of these two groups separately and in detail, he was then ready to turn attention to the arena in which they interacted.

You might introduce some important players, then put the scene in motion. Foster's approach offers a straightforward way to organize a study of directed change. An awful lot (both literally and figuratively!) of dissertation study and funded research has focused on the topic of directed change, although such studies typically lack adequate attention to the full context of it. In terms of sorting, one might begin with the very categories that Foster identified: Target Group, Innovating Organization, and Interaction Setting (Foster 1969: Chapters 4, 5, and 6, respectively). Those categories not only provide an excellent way to organize a study of change processes, they also provide a model for organizing any field study focused on the effects that those in one status want to produce on those in another.

To sort your data, begin with a few categories sufficiently broad to allow you to sort *all* the data. Remember that you are only sorting. If you are having problems with what ought to be a simple, straightforward task, you are probably beginning (or hoping) to develop theory, regardless of how modest. You are trying to take two steps at once. Try taking them one at a time!

I first encountered this dilemma as I began to organize the material from my huge (in terms of pages of notes) ethnography of a school principal. I kept adding more and more categories, and the sorting became increasingly complex. I had the good luck to meet Howard Becker at the time, and he suggested that if I was having trouble sorting things out, I must be doing more than sorting things out. Was I, perhaps, beginning to develop theory, a next step? Once I simplified the system, the sorting became easy. With my data adequately sorted, I could consider some refinements.

Becker's advice in those pre-word-processing days was to put important bits of data (quotes from informants, an observation, a vignette, an insight recorded during note taking) on individual 3×5 or 5×8 cards, or half-sheets of typing paper. The sorting was literal: one sat at a table (or on the floor) and physically sorted a stack of data "papers" by putting them into smaller piles according to categories that allowed a first run at the organizing task.

The big technological breakthrough of the day was the keysort punch card, an index card ringed with holes that could be punched open according to whatever coding system the researcher devised. With your keysort cards lined up, each card code-punched to index the data contained on it, you had only to insert one or more rods through the entire deck of cards and then give them a good shake. Your coding and punching system allowed the rods to separate the cards containing selected data from the rest.

I grew accustomed to putting data on 5×8 papers, easily typed or handwritten, easily stored, and easily sorted. My stacks of cards or papers must seem archaic in this computer age, but I describe them to help you visualize processes partially hidden by technology. I encourage students engaging in preliminary fieldwork exercises to do the same, manually manipulating actual bits of data—rather than electronic bytes—to get a physical feel for what they are trying to accomplish. Software programs make all this faster and less cumbersome today (for **further discussions** of coding and analysis, see Bernard 1994, 2000; Ryan and Bernard 2000; Weitzman 2000), but keep in mind the simple task you are trying to accomplish, and keep the task you are trying to accomplish simple.

I marvel at new programs available for data management and analysis with personal computers, particularly those programs that are especially suited to the requirements of qualitative researchers, such as text management.[4] But I share with others a concern that most programs are better attuned to the almost limitless capacities of the microcomputer than to the finite capacities of human researchers. For example, in THE ETHNOGRAPH, one of the earliest programs developed specifically to handle the kinds of data obtained through fieldwork, and still going strong,[5] a "line" of text—ranging from a single word to an entire interview—can be coded into as many as seven different categories. That was obviously a great convenience when one finally arrived at the categorizing stage, but it presented an almost irresistible temptation for anyone who had not yet reached that stage. For someone having trouble *sorting* data, a program that allows for no more than two or three categories proves a blessing in disguise. Moreover, computers are so engaging that they draw researchers away from the central task of thinking about their research focus and reorient them to a data-entering ritual that is often tangential to the research problem itself, a great opportunity for avoidance behavior and far more fun!

Again I note: The critical task in qualitative research is not to accumulate all the data you can, but to "can" (i.e., get rid of) much of the data you accumulate. That requires constant winnowing, including decisions about data not worth entering in the first place. The idea is to discover essences and then to reveal those essences with sufficient context, yet not become mired by trying to include everything that might possibly be described. Audiotapes, videotapes, and computer capabilities entreat us to do just the opposite; they have gargantuan appetites—and stomachs. Because we can accommodate ever-increasing quantities of data—mountains of it—we have to be careful not to get buried in avalanches of our own making.

This problem of data overload has an earlier parallel in one of the unanticipated consequences of the copy machines now a common feature of everyday life. In my student days, when we were given a reading assignment, we went to the library and read the material, typically distilling key ideas by copying in longhand a few telling quotes. Today's students copy (or download) entire articles—even whole books—often without going to the library at all. Their visits to their sources, whether real or virtual, are more efficient than ours were, except in one important way: at the end of the visit, they have not yet begun to satisfy the assignment. When they do get around to their reading, they will probably underline or highlight huge passages of text, rather than identify key phrases and summarize the rest in their own words. That is not good training for doing descriptive research. The parallel style in fieldwork is referred to disparagingly as the "vacuum-cleaner" approach: the fieldworker attempts to see and record *everything,* and that simply cannot be done.[6] There is a distinction to be made between researchers who try to impress us with how *much* they have observed and those who impress us with how *well* they have observed.

Let me bring this message home with the consequences of the vacuum-cleaner approach for a dissertation study that was being written while I was preparing the first edition of this monograph. I was a member of the dissertation committee, not the chair, but in some ways I felt responsible as the instructor who had "inspired" a capable doctoral student to embrace a qualitative approach on the basis of coursework taken with me. The student, Alfred (a pseudonym, of course), was a veteran public school teacher with particular interest in social studies. Alfred originally proposed an ambitious year-long study of his own social studies classroom, to include videotaping, diary-keeping, test results, student reactions, observations conducted by independent observers—you name it, he included it. I eventually realized that as members of his dissertation committee, we were remiss in not pressing him for a tighter focus in his proposal. We did not want to appear carping, so we lauded Alfred's thoroughness and capacity for detail when we might better have insisted on his doing less. Too, we already had at hand his massive review of the literature, which we mistakenly took as a good sign instead of a bad one, for it was exhaustive rather than selective. What a service a section on "delimitations" might have provided!

We didn't see Alfred on campus again for several years. He returned to classroom teaching, having used his allotted study leave to pursue coursework prior to doing his dissertation research rather than after it, the typical but unfortunate pattern of many doctoral programs in professional schools. We kept receiving progress reports that things were going well—for example, that the first three chapters were "already completed." Realistically, that news didn't reflect as much progress as it seems. His proposal had been so extensive that it virtually overwhelmed Chapter One. The literature review constituted a ready-made, if tedious, Chapter Two. Chapter Three exhausted method. At that point, Alfred became so mired in data that, like Anne Lamott's brother trying to prepare a classroom assignment about birds, he was immobilized, caught in the trap of inadequately focused research. The face-saving element was that his problem was attributed to writer's block. He did eventually finish, but without the sense of satisfaction he had anticipated. It had been more like a prolonged and unpleasant experience in accomplishing something minimally satisfactory instead of a highly anticipated first step into a new phase of his career.

SUMMING UP: PUTTING IDEAS TO THE TEST

Let me pause to check on whether you feel you can "get going" before I turn to a chapter intended to help you keep going.

- Try drafting (or revising) a "statement of purpose" for a possible study that you might conduct or a descriptive article you already have in mind. Use your statement as a way to make the ideas discussed here pertain specifically to your professional interests. Now, can you winnow your written statement of purpose to 25 words or less?

- What if you happen to feel that I have overestimated you? You do not feel that you have any grasp of how to proceed or even how to propose a statement of purpose. Proceeding with any kind of writing seems out of the question. There is still hope.

Forget all the steps suggested above, and simply tell the story of what you did and what "they" did. Answer the question, "What is going on here?" If there was some sort of event or "treatment" involved, tell how things looked before and after the treatment. Describe what you observed in careful detail. Because you are describing from one point of view only (your own), try describing the way you imagine it looked from the viewpoint of others at the scene. There, that gets you started, and with that much of a start, others might be summoned to help you couch the kind of issue you are trying to investigate.

- I assume, with Sholem Asch, that when you have something important to say, you can find a way to say it. Think of me looking over your shoulder, reading what you have typed onto the screen, perhaps pushing you with a prompt like, "In terms of what?" or "Then what happened?" or "Why are you telling us that?" Regardless of the explanation you offer, count on me to insist that you be more specific, more revealing, more . . . honest and direct? Don't worry if you are perplexed; no one will insist that you have to know everything!

- Once you have something written that suggests your purpose, try to refine (and perhaps shorten) it. Then, from what you know, identify the major topics that you plan to address and the order in which you intend to address them. Assign each topic a tentative number of pages. Then begin your writing, using the writing process to help you continually to refine your statement of purpose.

- Find some relatively easy place to begin the writing—perhaps describing how or why you approached the topic. You don't have to begin at the beginning. Write whatever parts seem ready to be written; you can plug in sections as they are completed. You don't need an introduction until you have something to introduce. If you find the idea attractive, consider writing a very rough first draft before you begin the actual research.

NOTES

1. In a workshop on ethnographic writing held in November 2000, anthropologist Regna Darnell suggested "standpoint" as an alternative to the sometimes ambiguous term "voice."

2. For instance, reporting percentage figures for the often tiny populations with which we work. Consider the difference in reporting that 23% of village households had washing machines or that 3 of the 13 households had them.

3. For an example and extended discussion on editing and using historical documents, see Stevens and Burg (1997); for internet inquiry see Markham and Baym (2009).

4. Text management is only one of the functions that software programs address. Weitzman identifies four additional software types: text retrievers, code-and-retrieve programs, code-based theory builders, and conceptual network builders (2000:808–810). The wide variety of programs serve different functions. Information on the web can be particularly helpful on this point. To learn what Wikipedia has to say about qualitative research, you might want to see http://en.wikipedia.org/wiki/Qualitative_research. See also Lewins (2009).

5. For the most recent updated information on THE ETHNOGRAPH, see http://www.qualisresearch.com/.

6. Howard Becker reports how he used to "drive students crazy" in his fieldwork seminars by insisting they report in greater and greater detail until they finally realized that it is *impossible* to observe and record "everything"; at that point, he wanted them to think about what they *really* wanted to learn from an observation, because "the trick in observing is to get curious about things you hadn't noticed before" (Becker 2007:88–89).

CHAPTER THREE

Keep Going

The precondition for writing well is being able to write badly and to write when you are not in the mood.

—*Peter Elbow,* Writing With Power

Once writing is under way, with something to say and a plan for saying it, you should make satisfactory progress on your own. Not only would you be better off left alone, I doubt that you would take time to read something like this, because it is addressed to a problem you do not have. Keep in mind that each individual writing assignment takes a unique direction. The best advice I have for any writer already writing is conveyed in my working chapter titles: Once you *Get Going,* then *Keep Going.* (I remind you that these are intended only as working titles. If I can make you conscious of the need for editing in *my* writing, maybe I can make you conscious of the value of editing in yours.)

In addition to being an act of arrogance, writing is a test of one's tolerance for delayed gratification. Even when the writing seems to be going well, there could hardly be gratification enough to warrant the time and commitment necessary to keep on keeping on. You work without feedback or encouragement. As Professor Aubrey Haan reminded me years ago from personal experience, writing is a labor of love. Your only measure of progress may be the diminishing number of subtopics still to be addressed and a slowly mounting stack of pages with text of uneven quality that probably falls short of your original aspirations. The number of pages may cause you concern, whether too few, raising doubts as to whether you are providing adequate detail and explanation, or more likely, too many, an indication that you may be creating a new problem. Eventually, you will have to do some cutting to stay within your intended page limits. Don't be distracted about length too early. For the present, just keep plugging away. Regardless of whether you are underwriting or overwriting, you *are* writing. As Charles Darwin wrote to his friend more than a century ago, "It is a beginning, and that is something."

I can suggest a few pointers that may help you to keep going, but the issues I raise are more important than the resolutions I propose. "Anything goes" that results in a tangible written product moving you toward a working draft that offers a toehold for subsequent editing. For analogy, let me suggest the criterion used by most of my Chinese acquaintances around the world whenever I ask if I am using chopsticks the proper way: "Harry, is the food getting to your mouth?"

In writing, results are what count; the end justifies the means. How much coffee you drink, sleep you lose, days you "waste," even how awful your first drafts look—none of these matters really matters. Be ever mindful of Becker's wise counsel, "The only version that counts is the last one" (1986:21).

STAY WITH IT

If you have engaged in substantial fieldwork, be prepared to spend from several months to a year or more to complete your writing. Rosalie Wax's sage advice was to allow as much time for analysis and writing as time spent in the field—and even more, if you are "really astute and can get away with it" (Wax 1971:45). From the outset, pace yourself for an activity in which it is critical to *sustain* interest, not merely to capture an occasional burst of energy of the sort that gets you through class assignments (term papers included). "The precondition for writing well," Peter Elbow observes in the epigraph quoted above, "is being able to write badly and to write when you are not in the mood" (1981:373). Set reasonable expectations, but be demanding of yourself. Think how you could churn out books if you wrote only one page every day! Make and keep this commitment: that in your daily or weekly schedule, the time you allocate for writing will receive top priority. I read that authors who make a living at it attend to their writing not only on a daily basis but for a seven- rather than a five-day week.

You understand that here I use the term "writing" in more than its literal sense of putting pen to paper or fingers to keyboard. Not every moment you devote to the preparation of a manuscript will result in the production of text. In spite of what I said earlier, writing covers a constellation of activities, including planning, organizing, and analyzing, as described in the previous chapter. Additional new demands arise as a manuscript begins to take shape: rereading, re-sorting, refining, rechecking, revising, and time for just staring into space ("ruminating," if you share my fondness for alliteration). Friends who try to be encouraging don't necessarily confer a favor with their incessant inquiries, "Well, how's the writing going?" Maybe you can cool them off with a reminder that writing entails more than simply putting words to paper, but my guess is that first you have to convince yourself.

THE "EMPTY FOLDERS" APPROACH

During the early 1970s, when qualitative approaches were really catching on, I served as an "outside consultant" for a nationally funded field-based study of educational change in which a number of qualitatively oriented researchers, all with backgrounds in anthropology, sociology, or educational research, were hired to conduct long-term studies in ten rural communities throughout the United States. A major responsibility for each of these resident researchers (on-site researchers, or OSRs, as they came to be known in project lingo) was to develop a monograph describing the community, the schools, and the nature and consequences of the effort at educational change that was the distinguishing criterion among the participating rural school systems.

From the outset, project directors of this independent and long-term evaluation effort were concerned that not every on-site researcher would actually complete the requisite monograph (for more on the project, see Sturges forthcoming). As an outsider, I was asked to think of ways that would foster success without infringing unnecessarily on the enthusiasm, independent spirit, and effort of each fieldworker. The directors also recognized that preparation of the case studies had to compete with numerous other responsibilities imposed on the researchers, such as gathering survey data for a cross-site study and continuing to do fieldwork throughout the duration of the project. Nor did the project want to impose a rigid schedule of deadlines that required everyone to submit manuscripts in lockstep fashion on a predetermined outline of topics.

As the directors correctly anticipated (and feared), it was long after the project ended before the last of these final reports was finally submitted. That eventually they all were completed reflects favorably on the integrity of the researchers: you cannot force people to write. You can tie remuneration to receipt of a completed report in funded research, but, given the long duration and multifaceted nature of this project, these researchers were on annual salaries. There was no way to threaten them by withholding funds without threatening the success of the project. Under such a threat, any researcher experiencing difficulty drafting the case study might have found an excuse to quit the project and escape from what was, for some, looming as an onerous task. Most had only recently completed their dissertations; the prospect of another major writing assignment "under pressure" generated unanticipated anxiety for several of them.

I recommended that each on-site researcher initiate the writing assignment by proposing a Table of Contents for the monograph to be prepared for that site. (Does that surprise you?) With that task completed, a collective review of the proposed Tables of Contents for all ten sites might constitute the agenda for a major project seminar. At that seminar, researchers could discuss, elaborate on, and defend their ideas. The project coordinators could also suggest (or impose) any standardization of format deemed necessary for the project as a whole. True, that might have precipitated some critical and delicate negotiation, but the research organization did have contractual obligations to meet, as well as a commitment to treat the fieldworkers as competent professionals. Following that negotiated agreement, the preparation of individual monographs would begin.

Next, I proposed that each fieldworker prepare a set of folders, one for each intended chapter of his or her projected case study, plus extra folders as needed for keeping track of miscellaneous materials, topics for local research, names of people to contact, possible bibliographic resources, and so forth. For the purpose, I thought a set of "hanging" folders would be ideal, so that not only computer printouts but other accumulating materials—original letters, photos, handwritten notes, even whole documents—could be dropped into the proper folder.[1] Eventually, of course, the relevant text material would be entered into a computer format, but the hanging file folders could continue to serve their repository function and provide a place for collecting supplementary materials as well as printouts of earlier drafts.

The contents of each folder would evolve through a roughly comparable sequence, beginning with brief memos or jottings or a set of data cards, progressing either to a tight outline for the chapter or a rough first draft, and thence, through revisions as necessary, to a completed draft of a chapter ready for inclusion in the evolving monograph. The problem of employee accountability, under circumstances that I dubbed "contract anthropology" (HFW 1975:110; see also Clinton 1975, 1976; Fitzsimmons 1975), was revealed in such "hypothetical" questions posed at headquarters as "How do we know whether the fieldworker is really at the site?" or "How do we know whether progress is being made on the monograph?" Those pervasive concerns could be alleviated simply by periodically asking each fieldworker to forward evidence of progress *in any one folder.* Meanwhile, fieldworkers would have wide latitude from one reporting period to the next in deciding whether to devote their current effort to preparing an outline for a proposed chapter, to writing a first draft of a new chapter, or to revising and refining earlier drafts of chapters as the anticipated monograph began to take shape.

Built into this production scheme was a recognition that no two researchers were likely to be, or needed to be, working on the same topic or at the same speed. It also allowed for periods when formal productivity might be low. At times, for example, efforts on the individual case studies were sidetracked by project-wide assignments or by attendance at professional meetings when the fieldworkers reported on their work in symposia directed at audiences of peers and patrons.

Although the procedure as described was never formally adopted for the project as a whole, I know it proved invaluable to some of the researchers individually, and I still think it was a very practical suggestion. I continue to tout it as eminently workable, as useful for a lone researcher as for someone coordinating a large-scale project involving parallel studies at multiple sites.[2] I realize that the folders that I envisioned could be created on the computer now, but the critical problem remains: how to ensure that everyone's work moves forward, however slowly, rather than allow it to come to a standstill.[3] The idea is something of a writer's adaptation of the first law of motion: Authors with a manuscript in motion will keep it in motion, authors with a manuscript at rest. . . .

In addition to place holders assigned for each major chapter, one's set of folders for a work in progress—whether the "folders" themselves are real or only imagined—ought to include a place for anticipated short assignments like preparing a draft of the acknowledgments or updating the list of references. The full set of folders for a project might also include proposed symposia or seminar papers, as well as articles intended for separate journal publication. I emphasize the idea of researchers with chapters at various stages of development.

KEEPING UP THE MOMENTUM

A major writing project such as a monograph or thesis does not proceed with every section at the same stage of development. The more ambitious the total project, the more advantageous to have different sections at different stages of development, so that chores can be varied, and time and mood accommodated. Unforeseen delays should bring neither the research nor the writing to an abrupt halt. Anticipate (and expect) delay and be prepared to turn to other tasks, perhaps even the preparation of the first draft of your *next* article, proposal, or project. Hard to imagine just now, but there may even come a day when you can do this kind of scheduling with a number of "irons in the fire." If you have writing tasks at several stages of development, you can remain productive in spite of delays in the review process or production schedules.

Opinion varies as to which stage is hardest. In my experience, the first draft of anything I write is always the most difficult one. Provided that I am off to an adequate start, I find pleasure in feeling that my manuscript is taking shape through the subsequent revising and editing, even when the increments are small. No question that revising and editing are critical tasks. To some, these tasks are the most difficult, but I do not concur with Peter Elbow that they are the most unpleasant (Elbow 1981:121). For me, writing enervates and editing exhilarates. The only unpleasant feature about editing is in acknowledging how awful some of my sentences are as originally written. (I started to collect examples of some of my worst sentences but decided that I did not need to convince you that I am as capable of writing them as anyone.)

I cherish the advice recalled by Denise Crockett while she was struggling with her dissertation: "If you can't write well, write shittily." You have to have *something* written before you can begin to improve it. In *Bird by Bird,* author Anne Lamott not only recognizes the *possibility* of writing "shitty first drafts" but insists that most writers begin with them: "The only way I can get anything written at all is to write really, really shitty first drafts." But that does not bother her. She consoles, "All good writers write them. This is how they end up with good second drafts and terrific third drafts" (Lamott 1994:21–22).

Sometimes the writing goes excruciatingly slowly. On days when it doesn't seem to be going at all, you might devote some time to bringing the reference section up to date. That leaves you armed with a ready reply, should some insensitive but well-meaning colleague raise the anxiety-provoking question, "Well, how did it go today?"

A suggestion that experienced writers offer in order to regain momentum when you return to your writing again (i.e., tomorrow) is to pay close attention to where you decide to stop as you come to the end of the day's writing session. The advice is to stop at a point where you know you can easily start up again. At the least, jot some key words that capture your train of thought. If you are in the middle of a paragraph that you know you can finish, stop there. If you are copying a long quote from an academic source or an informant, stop at the *beginning* rather than at the end, so that when you start again you can get right to work.

(In actual practice, however, I usually do just the opposite. You probably do, too. I stop when I am stuck and return somewhat hesitantly to see if I can work my way out of the mire the next time— thus the old adage, "Do as I say, not as I do." And I often begin my day's writing by reviewing and editing what I wrote the previous day. Admittedly, that is a slow way to get a fast start, although the advice to begin by reviewing what you were writing the previous day is also heard frequently. I think I begin that way because I find editing more satisfying than writing the first draft.)

Editing obviously can become an escape from writing, or at least a hindrance to getting through a first draft. On days when the sentences do not flow, looking back over yesterday's work does offer a way to get warmed up. Having struggled with particular words or ideas on an earlier attempt, I sometimes see a better resolution on my next try. The editing-reviewing may take up to an hour—about one quarter of the minimum time I try to set aside for writing. It also violates my Puritan ethic, which holds that the pleasure (editing) should come *after* the pain (writing), not before. But it is a concession I make in order to accomplish my major objective: to keep at it, once the writing begins. Try to make some measurable progress in the development of your manuscript every writing day.

WHEN IT'S TIME FOR DETAILS, GET THEM RIGHT THE FIRST TIME

The proper form for citations, references, footnotes, margin headings, and so forth required by your discipline, your institution (if writing a thesis or dissertation), or your intended journal or publisher should be clearly in mind as you work. Your default mode should be the accepted standard for your discipline, a style with which you need to be thoroughly familiar. When preparing material for publication in a format with which you are not familiar, have at hand a recent issue of the journal, an authoritative style manual, or the web page address for access to the journal's style manual.

You may think it unimportant to bother with such picayune detail as proper citation form in your early drafts. "First things first," you rationalize; why worry about little details until you have some text in place? That might be true if you are tempted to check every source or hunt down every quotation when you first introduce it. Better to push on, concentrating on the gist of what you are writing rather than getting bogged down in detail. But it is easy to note details that need checking, perhaps by marking them in some special way for attention (e.g., with **boldface** or underlining). I assure you that your time will be well spent if, at whatever point you do attend to details, you do so carefully, correctly, and fully, in the form in which the piece is to be submitted. The earlier you get these things recorded correctly, the better.

In the old days, there was always the likelihood of introducing new errors into previously correct copy every time a manuscript had to be retyped. A comforting aspect of working with the computer is that once you get something written—barring rare technical glitches—it is going to stay that way. So get it right the first time.

The more details you attend to in the early stages, the more you can direct your attention to content as the writing progresses. You also free yourself from having to look after such details during final revision(s) when your attention should be on whether you have your words and ideas in a proper sequence, not worrying about someone else's. When feasible, I recommend that you retain a copy of any material that you might later want to quote at length, regardless of how remote the possibility. That way you can quickly double-check or respond to a copyeditor's last-minute query, "You sure it was *exactly* like this in the original?" When I may want to quote printed material that carries over to a second page, I also note where the page break occurs. Should I later decide to use only a portion rather than the entire quote, I know how to cite it without having to go back to the original to check pagination.

Developing a Style Sheet

Are you aware of the variation in the spellings offered in different dictionaries, the citation forms preferred in different fields (as well as preferences from one journal to the next in the same field), and the options about the form and placement of footnotes or endnotes? When you do become aware of such niceties, you will be amazed both at the number of decisions that need to be made (for example, in capitalization, hyphenation, use of the serial comma) and at the extent of indecision that surrounds certain choices as to preference in style. An example from recent experience is the phrase "participant observation." The flagship journal of the American Anthropological Association, *American Anthropologist,* has its style sheet available on a web page and it shows *participant-observation* as a hyphenated phrase. Yet the *Anthropologist* treats the phrase as two words. For you as budding author, this ambivalence is both bad news and good. The bad news is that there is no ultimate authority in language usage for English. The good news is that in cases where no one seems to be in charge, you can take charge yourself, at least to some extent, in writing a book. Here's how.

If you have never published, you may not realize that copyeditors develop an **individual style sheet** for *each* book-length manuscript. That style sheet provides a record of all decisions pertaining to your specific manuscript that are not already covered by an existing style sheet for that publisher. There is usually a current style sheet for major journals as well. Even if there is nothing in print, *somebody* in an editor's office exercises final authority on all decisions that are not left to the author. A style sheet records the decisions about spelling, hyphens, commas, formats for headings and subheadings, footnotes, and anything else that needs attending to, in order to ensure that usages within the text are consistent and the overall text is consistent with the publisher's preferences. If you are writing an academic thesis, you will discover that your graduate school (or some comparable office) has assumed responsibility for this function, so it, too, has a style sheet, the institution's final opportunity to impose its authority. Prepare yourself for some firsthand experience with institutional rigidity should you deviate from its so-called guidelines.

It is a good idea to develop your own style sheet for a manuscript, even if it consists of nothing more than a sheet of paper with your decisions (or indecisions) about spelling, hyphens, and capitals. Keeping a style sheet encourages you to track troublesome words as you become aware of them (e.g., adviser or advisor, gaining entry or gaining entrée, judgment or judgement, macro-culture or macroculture: which form are you going to use?). Your style sheet may not guarantee your authority in any decisions to be made, but at the least it can help you to identify inconsistencies, to alert copyeditors that certain (often reappearing) terms are causing you problems, or to remind you to check with the graduate school as to local "preferences."

Style sheets do reflect preferences and customary usages. Be prepared to capitulate if you find yourself at odds with editorial or institutional policy, but don't give up prematurely. Styles are always in flux. Publishers' style sheets and the major style manuals are constantly being revised. The authoritative *Chicago Manual of Style* is well into its "teens" in revisions.

Keeping Track of References

In similar fashion, follow a consistent style for maintaining a personal file of bibliographic references. The obvious choice should be the standard in your field (if one exists) or the style of one of its major journals. You may want more detailed information than is ordinarily required by any of the abbreviated formats:

- full and complete title and subtitle of every source cited;
- *full* names of authors and editors (i.e., not just first initials);
- full journal names, with volume and issue number;
- inclusive page numbers for articles and for chapters in edited volumes;

- publisher's full name; city and state where published; and
- date of the publication you consulted, as well as the original date of publication, if different.

Not all of this information is required by every journal or publishing house. Yet it takes only a moment to make a complete record in your original notes, and it can save time if you should discover, for example, that a journal to which you have submitted your article uses authors' full names rather than only first initials.

When citing material published long ago but accessed by you in a more recent edition, be sure to give the date of original publication as well as the date and page of the edition from which you are quoting (e.g., Wolcott 1989[1967]:107). That way it won't look as though an author has sprung back to life or is passing off as new something written years earlier. Most journals provide illustrative entries for the way they wish citations to be formatted. They also provide instructions for citing electronic publications.[4]

Whenever possible, I also track sources *forward* by including in my notes and in formal citations any available information about materials republished or reissued. This practice is especially helpful for references to journal articles subsequently reprinted in books, or to previously out-of-print sources that become available again, as with many of the case studies in cultural anthropology and in anthropology and education originally published in the 1960s and 1970s.

One further suggestion about academic references: Make your citations as explicit as your text warrants. There are occasions when a reference to an *entire* work is appropriate, although if you cast a critical eye over the way academic writers parade their citations, you will catch some of them in a shameless game of name dropping. They lob references like so many snowballs over a fence, an indiscriminate barrage that fails to achieve the kind of specificity appropriate in scholarly writing. To be really helpful, go beyond minimum expectations (author's last name and date of publication) to inform your readers of the exact page number and the nature of the material to which you make reference, and, unless it is apparent in the text, your reason for citing it (i.e., whether it is your source: "see"; a source of additional information: "see also"; or a source for comparison or contrast: "cf."). Most readers will not consult your sources; they count on you to inform them. That is one reason for being accurate and complete. Conversely, some readers *will* consult your citations. That is the other.

Keeping Track of Bits and Pieces

As a manuscript evolves, you might find it handy to keep track of possible topics or references to include, paragraphs deleted in one place that may fit better somewhere else, and so forth. For each developing manuscript, I maintain a separate document or file where I can park such "working notes" temporarily until I decide their fate. As memory fades, I have found it essential to do all such tracking on paper or screen rather than trust that I will recall those details when needed. I have also grown more cautious about the way I make even minor revisions of text. Rather than delete and then rewrite material, I now move existing material ahead a few spaces (usually by hitting "return" a couple of times), insert my rewrite, and *only then* delete the old if the text is really improved.

Over the years, I have also developed the habit of keeping a permanent set of brief passages, theoretical notions, aphorisms, possible chapter epigraphs, frequently seen foreign phrases, and well-stated ideas or advice heard or read. These I keep in a handwritten journal, although they can easily be kept on the computer, logged in as they come to my attention. My notebook is labeled *Quotes*. Sometimes these sayings are incorrectly or inadequately referenced when I discover them. If I am unable to track the original source, at least I can acknowledge the author. My journal of *Quotes* has proven a valuable repository and resource for ideas and pithy sayings.

GETTING FEEDBACK

The compound word "**feedback**" contains two elements. The first implies nurturance. Most authors crave it. The second indicates direction: turning back. Feedback draws attention to the already-done rather than the yet-to-do. Keep that in mind when you begin to long for it. Don't seek it too soon, especially if it might divert attention from completing the full draft by tempting you to start revising what you have already written.

I recognize the good intentions of professors who want to approve (which, unfortunately, may also mean disapprove) the first three chapters of their students' dissertations, but the advice I gave my doctoral students is the same I give to all writers: Work independently as long as possible, even including a draft of your tentative ending, before inviting feedback. When you are ready, seek feedback judiciously. A little goes a long way.

Timely and useful feedback on early drafts is hard to give and even harder to take. The problem is compounded in qualitative research because there are so many facets on which feedback can be offered: whether one has identified the right story to tell, how adequately it has been described, how well it has been analyzed and interpreted. As with any writing, it is also far easier for your reviewers to identify problems, awkward sentences, and alternative explanations than to know what to say about particularly well-conceived studies, particularly well-turned phrases, or particularly insightful interpretations, other than a clichéd, "I really liked this," or "Great!" Even the most gracious and gentle among your reviewer-critics are far more likely to fault weaknesses in a manuscript than to applaud strengths, unless they render only a global reaction and leave the nitty-gritty to others. Regardless of intent, feedback tends to be disproportionately critical and negative. Your consolation may be that the more painstaking the critique, the more you may assume that your critics have regarded your effort seriously.[5]

Choose early reviewers with care and instruct them carefully as to the kind (and extent) of criticism you feel will be most helpful at each stage. Unless I am developing a manuscript that has been solicited by an editor, I prefer to invite friends and/or fellow authors to be early reviewers. My assumption is that they constitute a support group who will look for ways to help me say what I am trying to say in specific instances, rather than dwell on my (or my manuscript's) apparently not-yet-attained potential.

Yet I value all feedback short of flat-out rejection. I would not think of formally submitting a manuscript that had not been given a critical once-over by colleagues, both as it was being developed and in almost-final form. I say "almost final" because as long as we invite critique, we will get it: the process never ends. If you insist on receiving final approval for something you have written, you will have to be candid about soliciting it.

In seeking feedback, keep in mind a distinction between the conduct of research and the reporting of research. Research purposes come first. Eloquence can enhance a good study, but it cannot rescue a poor one. Early readers should be directed to look primarily at the accuracy and adequacy of detail; at how the problem is stated as the account unfolds; and at the appropriateness of the description, analysis, and interpretation. Outside readers may recognize aspects of a study to which an otherwise preoccupied researcher has become oblivious. There may be little point in worrying about the niceties of style if the content is not in place, interpretation misses the mark, the focus is misplaced, or the account lacks balance. Also recognize that no reviewer is likely to have something to say about every aspect of your work. Steel yourself for the likelihood that, regardless of how you instruct them, your reviewers invariably will say more about style than content. It is, after all, *your* account. Others should see their role as helping you to convey *your* ideas, not to make you a vehicle for presenting their own.

An ideal combination of early reviewers might include a colleague from one's academic field, to attend to framework and analysis, and a reader familiar with the context or setting who reads for accuracy, completeness, and sensitivity to those being described. If you are able to cajole any of your earlier readers to read a later version, help them to help you by calling attention to sections that have been rewritten or added. Most reviewers are capable of only one critical reading, especially without some direction from the author. If you have served in the role of editor or director of dissertations, you are well aware of the difficulty of bringing a fresh perspective to multiple readings of a manuscript.

And what to do with the advice and suggestions you do receive? Your first obligation is to listen attentively. Don't argue, don't explain, don't get defensive. Take the advice under advisement, show your appreciation, and make sure that you understand anything that your critics tell you that they did not understand. Even if your critics are in a position to assert their authority, you may be able to negotiate a compromise. But never simply assume that you alone are being denied an essential freedom and that everybody else is free to write whatever they please. As the old saw has it, freedom of the press is reserved for those who own one.

Like many fieldworkers, I make an effort to invite readers in the setting to look at developing drafts (especially the descriptive sections). I regard that as an integral element of fieldwork, and I like to note in subsequent drafts any reactions and comments prompted by earlier readings. (For some pros and cons of this practice, see Emerson and Pollner 1988.) Today's informants and collaborators not only *can* but *do* read what we write. If you have not thought about that aspect of feedback, you might review the lessons in

Caroline Brettell's edited collection *When They Read What We Write* (1993). Furthermore, those among whom you study may wish to, or may insist on, reading drafts prior to their general circulation.

It is advisable to anticipate how disagreements—or sometimes just "unhappinesses"—are to be negotiated. My practice has been to offer to share pre-release drafts with interested informants and to inform them that I will *take under consideration* any reservations they express. I think one is ill advised to offer full veto power, even to key informants or anyone with whom you are writing a personal life history. If someone holds that power, your project remains in jeopardy throughout its entire duration. Researchers, too, are human subjects who need protection from unnecessary risk.

Let me repeat: I have always delayed sharing a developing manuscript for as long as possible. I want to be sure I've said what I want to say, and have tried to say it well enough that my ideas are clear, before subjecting my words to the scrutiny of others. During the academic year I devoted to writing my doctoral dissertation (following a full 12 months of fieldwork), I deliberately lived away from the Stanford campus and made brief visits only when I needed to use the library. I did not need the company of other anxiety-ridden dissertation writers to get my own writing done. I had a story to tell. I was determined that, should the initial draft prove satisfactory to only myself, I first needed to recount the story my way. I sought little advice from my dissertation committee prior to submitting a completed draft to them. Had that draft been unacceptable, I was prepared to undertake whatever rewriting was necessary, but not until I had made my own version a matter of record.

I'm happy to report that except for reservations about length, and some useful editorial suggestions, the thesis was accepted as submitted. Little doubt that having one's thesis accepted without hassle can prove a great incentive toward further academic writing! Although the expected audience for your thesis may be small, don't lose sight of the importance of the thesis to your career, *especially* if you intend to pursue further qualitative research in which you expect writing to play an important role.

Although I avoided premature "official" feedback during that period of angst and authorship, I eagerly anticipated long work sessions with my fellow graduate student Ron Rohner and his wife Evelyn. We met regularly to discuss our progress, exchange information, and share and critique drafts of our developing chapters. Our independent but somewhat complementary studies were based on anthropologically oriented fieldwork conducted at the same time, in neighboring villages, among the same people, and with the encouragement of Professor George Spindler, our mutual mentor (Rohner and Rohner 1970; HFW 1967). Like the fieldwork on which it was based, our writing proceeded in a climate of mutual help and support.

When time is of the essence, or you find yourself unduly concerned about how the writing is going, I recommend finding some patient soul (for that reason alone this probably will not be an academic colleague) who will read and provide intentionally encouraging feedback. Better still, ask someone to read your words aloud to you, perhaps even to read without comment or with only general and supportive suggestions, such as "That reads well" or "This needs more explanation." Hearing your words read aloud can help you concentrate on what has actually reached paper, the experience you are creating for others out of experience that was originally yours alone. They are not the same.

Another reason for hearing your words read aloud is that we do not recognize the rhythms and patterns of our own speech. What we write usually reads well to us (i.e., literally "sounds right") because we know how to read it. But no two humans share identical patterns of speech or intonation. When that oral reader stumbles—or not-so-subtly gasps for air, as my dear friend Anna Kohner used to do while reading aloud the longer sentences of my dissertation drafts—the author needs to get busy with the red pencil.

Technology is exerting its influence on editorial practice as on every other aspect of writing and publishing, and you may need to adjust the match between the editing help you seek and the extent of help given. I can appreciate technology that facilitates team review of a collective document or allows a newsroom editor to make changes directly on copy as submitted by a reporter. Because it is now possible for reviewers to insert "comments" without actually making changes in the document being reviewed, that is the mode that appeals to me. I cannot imagine burrowing into someone else's document to install changes that, particularly in the initial stages of a manuscript, are meant only as collegial suggestions.

SUMMING UP: TIPS TO KEEP YOU GOING

Let me conclude this chapter by reiterating the central idea: Keep the writing moving forward. Get the essence of your study committed to paper, no matter how rough or incomplete it may seem. Do not lose sight of the fact that well-focused interpretive statements may help you improve the problem statement, just as your developing analysis may help you make better decisions about the descriptive material, although the descriptive material will probably (but not necessarily) precede it in the completed manuscript. Further thoughts:

- Keep your focus in mind as you weave your story and your interpretation, but maintain a healthy skepticism about the focus itself. Always consider the possibility that you are not yet on target or that the focus has shifted in the course of your inquiry. A guiding question: "What is this [really] a study of?"

- Your major concern, especially in writing the first draft, is not only to get something down but also to get rid of data—to focus progressively, to "home in" on your topic. Keep track of tangential issues that you might (or should?) leave for another time.

- Do not allow yourself to get stuck because of data you do not have or problems and elements that you do not fully understand or cannot interpret adequately. Make note of whatever is bothering you, either for yourself, if you think things can be remedied, or for your reader, if the problem seems likely to remain fixed at that stage. Then get on with it. Readers will not be offended if you do not presume to know everything.

- Unless absolutely forbidden to do so by a stuffy editor or dissertation committee, write in the **first person**. Put yourself squarely in the scene, but don't take center stage. The world does not need more sentences of the sort that begin, "It appears to this writer . . . ," or "What is being said here is. . . ."

- Try writing your descriptive passages entirely in the **past tense** if you find yourself moving uneasily between present and past. Admittedly, the past tense seems to "kill off" everyone as soon as an action is completed. It does strange things to "alive and well" informants, particularly if you begin writing while still in the field. By the time your manuscript has gone through several iterations, editorial review, and quite possibly publication, you will discover that the past tense no longer seems so strange. Nor will you have left informants forever doing and saying whatever they happened to be doing and saying when you last saw them.

- Use your extensive field notes and fieldwork experience to provide **concrete examples** and illustrations. Never underestimate the power of specific instances to support your generalizations—not simply to inform, but figuratively to reach out to your readers. Clifford Geertz challenges us to use "the power of the scientific imagination to bring us into touch with the lives of strangers" (1973:16).

- **Write for your peers**. Pitch the level of your discussion to an audience of readers whom you assume to be deeply interested in finding out what you have been up to. Write your dissertation with fellow graduate students in mind, not your learned committee members. Address your subsequent studies to the many who do not know, rather than the few who do. Editor Mitch Allen cautions against the academic tendency to write for what he calls one's "WNC" (Worst Nightmare Critic), the individual who knows more than you do and cannot wait to pick you apart. As Mitch observes, these critics are not your audience and "they will probably trash you anyhow." Don't cater to them.

- Give emphasis to important points you develop. Where we mean to write seamless prose, the result is often merely uninterrupted prose. Give ideas some room by being attentive to paragraphing. Make generous use of headings and subheadings to call the reader's attention and to mark shifts in focus.

- Heed the admonitions so frequently heard in the interest of better writing. Avoid wordiness, passive or convoluted constructions, long words and pompous phrases, abstract nouns and faulty pronoun references, misplaced modifiers, and nonparallel constructions. But don't allow such admonitions to hinder initial efforts to get your ideas written down. You can attend to style and correctness in the later stages of revising and editing, and you can get—and even buy—help from others with those aspects of writing. No one will ever see your early drafts. As your ideas take shape and become more elegant, take pleasure in crafting sentences worthy of them.

- Hold off on seeking feedback until you yourself have taken your study as far as you can go. Do not seek help that is premature or that you do not intend to use—capture your own ideas first before involving others.

NOTES

1. Thus, the files are sometimes referred to as "drop files."
2. For instance, a colleague with an educator's interest in the events of a particularly eventful year in American public education and a historian's

penchant for collecting data amassed so much data that he became immo-bilized with how to sort it and where to begin. I suggested assigning one folder to the events of each month of that year and then to developing the account one month at a time. No particular need even to address them in order; one might start with an "easy" or especially interesting month. Such a "bird-by-bird" approach did not capture my colleague's fancy; the study was never written.

3. Procedures for cross-site analysis received major attention in both editions of Miles and Huberman's *Qualitative Data Analysis* (1984, 1994); see also Noblit and Hare (1988). Such issues get into problems of synthe-sizing and aggregating cases, which are beyond the scope of this monograph.

4. Two examples of electronic reviews are provided in the Reference section under Wolcott (see 1999b, 2008b). The version in my References for 2008b varies from the one suggested by *TC Record,* which I cite here in full:

Wolcott, H. (2008). Telling about society [Review of the book *Telling About Society*]. *Teachers College Record.* Retrieved from http://www.tcrecord.org/content.asp?contentid=14871.

Clearly, there is as yet no standard form for electronic reviews. For more on APA style for electronic references, see http://www.apastyle.org/elecref.htm.

5. C. Deborah Laughton, editor of the second edition, caught me off guard with an opposite and unfamiliar tactic. She peppered my margins with laudatory comments ("Lovely," "Key point," "Nice rhythm," "True") so gen-erously bestowed that I felt I should make a critical examination of every page that failed to earn an accolade to see what I might do to bring it up to snuff. Don't count on finding many academic editors who follow this approach.

CHAPTER FOUR

Linking Up

Man is an animal suspended in webs of significance he himself has spun . . .

—Attributed to Max Weber by Clifford Geertz (1973: 5)

To this point, I have focused single-mindedly on the stated purpose of your research and have urged you to do the same. I have gone so far as to suggest that you draw attention to a sentence that begins, "The purpose of this research is. . . ." You won't go wrong if those very words appear in your final draft and you make them sentence one of paragraph one of chapter one. Although that sentence is a rather unimaginative way to announce your purpose and begin an account, it should convey to readers what you have been up to. It may also keep *you* on course.

But research is embedded in social contexts and, like all human action, is *overdetermined,* the consequence of a multiplicity of factors. A preoccupation with focus directs attention narrowly inward to ask what the researcher intended and what has come about as a result. Researchers themselves—humans suspended in webs of significance of their own making—have contexts and purposes far beyond the immediate scope of their studies. Time now to expand the gaze, to look at research as the social act that it is and to the *multiple* purposes (note the plural) we seek in pursuing it in our professional calling. How do we link up our research—and ourselves—with others?

I draw particular attention to three topics that offer opportunities for linking up with the work of others. The first is the traditional **review of the literature**. The second is the paean to **theory**. Third is the customary discourse on **method**. The three topics have become so much a part of the reporting ritual that, in many qualitative (and most quantitative) dissertations, each may be assigned a separate chapter. Too often, the topics are addressed in elaborate detail before the reader catches more than a glimpse of what the researcher intends to report. Because old habits die hard, this rather standard "dissertation format" tends to reappear in writing subsequent to the dissertation, simply because authors continue to follow their same patterns.

Rather than underscore the important role played by each of these three dimensions in the research process *writ large,* here I want to explore some alternative ways for linking up with "the literature," with theory, and with method that complement and augment the research that *you* are reporting. To me, that seems preferable to regarding the three as hurdles to overcome or rituals to be performed before you are free to strike out on your own. But you must gauge your situation, the prevailing norms in your academic specialization, and, if you are preparing a thesis or dissertation, in your department as well. In the latter case, if institutional constraints are strong, or your committee members include faculty yet to be convinced about a qualitative approach, you may decide that a far, far better thing to do is to comply with the expectations set before you. Before you begin to rock the boat, be sure you are in it.

Do, however, be sure that the traditions that you are going to honor really exist and are not just part of the mythology surrounding dissertation writing or getting an article accepted for publication. I recall a discussion with a senior faculty member who insisted that her dissertation advisees prepare a lengthy Chapter Two reviewing "the literature." She defended her staunchly held position on the grounds that a review was *required* by our graduate school. I did not deny for a minute that she could insist that students preparing a dissertation under her direction include such a chapter in their completed studies (just as Alfred, who happened to be her advisee, had done). But I insisted that the "rule" was hers. I offered to accompany her to the graduate school to prove my point.

She allowed (privately) that the "rule" might not *actually* exist, but she **personally** insisted on such a review as evidence of her students' competence—and here she might have used the term "command"—of their field.[1] I had, and have, no argument with finding ways to have students demonstrate their newly won mastery of some special body of literature. But it has always seemed counterproductive to burden a dissertation with a secondary task diametrically opposed to demonstrating one's ability to focus on a particular phenomenon studied in depth. A command of the literature can be assessed through other preparatory assignments, as, for example, a separate synthesis paper included as part of the requirements in a graduate program.

What I propose is that instead of treating these linking activities as independent exercises—in a dissertation or any scholarly writing— you remain resolutely selective about the links that you make and you make relevant links on a when-and-as-needed basis. Most likely that will mean holding off except for the most general of comments until the research you are reporting is ready to be situated in broader contexts.

I am assuming, of course, that you have something significant to report. Such is not always the case, especially in dissertation writing. Realistically, what you have to report may be inconclusive, or what you have found points to a better restatement of the problem. If that is the situation, then a review and (brilliant!) synthesis of the work of others may become the critical mass to be showcased, the dissertation restructured (and retitled) accordingly. Or, your contribution may be a modest case study or the description of a new technique in fieldwork, whether successful or not.

I assume that the researcher does have plenty (too much, typically) to report and to analyze, which is often the case in descriptively oriented research. In such circumstances, one should not be expected to present a major review of everything that everyone else has done before reporting some original observations of one's own.

THE "LIT REVIEW"

Perhaps you paid close attention, even breathed a sigh of relief, when I suggested earlier that you dispense with devoting Chapter Two (which I treat here as a proper noun because the phrase has come to have its own special meaning) to a traditional literature review, especially if you are feeling, as Becker puts it, "terrorized by the literature" rather than aware of how to use it (see Becker 1986:

Ch. 8), and have yet to face that onerous task. Now hear the full message, not just the words you may have rejoiced to hear.

Let me remind you that what I tell you—in this chapter or anywhere in these pages—has absolutely no authority behind it. I am not one of the people who must be satisfied with your study. Citing me as an authoritative source for deviating from tradition is more likely to get both of us in trouble than to get you out of an obligation. If you are directed to write a traditional Chapter Two or its equivalent—a chapter or section in which you dutifully review the literature and/or outline your theoretical dispositions—by someone who *does* have authority, then do it you must. Perhaps you can negotiate the alternative that I propose. If not, accept the fact and rise to the challenge. Whether or not the experience will be "good for you" is difficult to ascertain, but I can assure that it could be bad for you if you do not. Note also that if you are asked to prepare such a chapter, it will be left to you to figure out just which *literatures* (note the plural again) you are expected to include—method, theory, prior research, social significance of the problem, philosophical underpinnings of inquiry, implications for policy, applications to practice, and so on.

My feeling is that readers want to be and ought to be engaged immediately with a sense of the problem you are addressing, rather than first be subjected to a testimonial demonstrating how learned you have become. They will assume you have a rationale for undertaking your research and will reveal it in time. They are not likely to insist that you plow through the entire history of research in your field before you dare take a step of your own. Readers come to their task ready to join you in a search for what you have to contribute. One of the things that makes *all* academic teaching and writing so boring—with dissertations topping the list—is the practice of approaching every topic with a backward look at where and how it all began.[2] Origins are important, but things don't necessarily need to be presented in the order in which they happened. A brief explanation about the significance of the topic should be enough for starters.

My proposed alternative to devoting an entire chapter to examining the underpinnings of your inquiry is that, other than presenting a brief justification for your study, you draw on the relevant work of others on a **"when-and-as-needed"** basis. (When-and-as-needed can serve as mantra for this whole chapter.) Such detail seems as apt to come *after* the presentation of new research rather than in anticipation of it. I object to the practice of simply backing up with a truckload of stuff and dumping it on unsuspecting readers, which seems to be what most traditional reviews accomplish. That is more likely to create an obstacle that gets in the way of, instead of paving the way to, reporting what you have to contribute.

Given the number of years in which you undoubtedly have been subjected to such an approach, you may feel duty-bound to follow it yourself. Well and good if you can weave your review into an engaging account without losing your readers along the way, especially if you are revising a dissertation for publication. But if the urge and urgency to provide a traditional review reflects the wishes of a dissertation committee, perhaps you can negotiate that the review be incorporated into your research *proposal* rather than into the final account. In that way you can demonstrate your command of the literature without having to force it into a predetermined place in the dissertation. In subsequent writing, if you feel a need to document how things came to be, draft such a statement and then set it aside. You can decide later whether your readers are likely to feel the same need as you did for such thorough grounding. By all means, consider looking for alternative ways to satisfy the *intent* of the literature review in your post-dissertation writing, rather than simply adhering to the same pattern because that is how you have "always" written.

Not for a moment am I suggesting that you can ignore the prior work(s) relevant to your research. Such thoroughness might be reserved for writing devoted exclusively to synthesis and critique, a form of scholarly contribution of great service to researchers in any field. Flag important citations to the work of others. But do so sparingly and only as the references are critical in helping you to analyze and to situate *your* problem and *your* research within some broader context. In the normal course of things, the need for locating your work within a widening circle of scholars is most likely to be toward the end of your discussion, as you draw the strands of the study together and ponder some broader implications.

In conducting a thorough review of the relevant literature, you cannot hope to get by with the pat response to investigating any question today: "Just 'Google' it." Google may be a great way to start a search, but you will certainly need to go deeper than that. You must identify who is doing or has recently done important work in the field you are investigating and/or how extensive their efforts have been to date, thus to secure a place for yourself and for the contribution you intend to make. Peruse the journals in which relevant topics appear, especially any theme issues, and follow up with relevant findings until you feel you have exhausted important leads and are literally a master of the topic.[3] But when you are ready to draw on your great fund of knowledge, be selective. Save the best and hold the rest.

MAKING THE LINK TO THEORY

In addition to reviewing relevant research conducted on the topic, you may be expected—or directed—to say something explicit about the issue of theory. No one will let you (or me) get away with the idea that there are no theoretical implications in our work, but issues of theory can be addressed in various ways. Theory should not be regarded as just another ritual to attend to, another obstacle along the route to obtaining a degree or getting something published.

I have suggested that you hold off on the lit review until the material you are introducing is well in place. Even more emphatically, I urge you to hold off introducing theory until it is quite clear what you are interested in theorizing about and how that relates directly to what you have to report. From the outset, resist the temptation to interrupt with premature excursions into analysis or interpretation, other than marking points to which you intend to return. This is not to suggest that the lines between description, analysis, and interpretation are so clearly drawn, but only that you keep the *focus* on the descriptive task until you have provided a solid basis for analysis and for determining where and how much to draw on the work of others.

When you are ready to address matters of analysis and interpretation, consider proposing *multiple* plausible interpretations rather than pressing single-mindedly for a particularly inviting one. We need to guard against the temptation to offer satisfying, simple, single-cause explanations that too facilely appear to *solve* the problems we pose. Human behavior is complexly motivated. Our interpretations should mirror that complexity rather than suggest that we have the capacity to infer "real" meanings. Qualitative researchers should reveal and revel in complexity, striving, as anthropologist Charles Frake (1977) has suggested, to make things appropriately complex without rendering them more opaque. Leave for more quantitatively oriented endeavors efforts to tie things up in neat bundles. They are better situated to do that, for, as Denzin and Lincoln observed,

"Quantitative researchers abstract from this world and seldom study it directly" (2000:10).

Interpretive remarks belong in the summation of your work where you situate your study in broader context. That is the place to draw upon the work and thinking of others. Scholarly writers sometimes succumb to a temptation that Raymond Firth described as making a "parade" of social theory (noted in Sanjek 1999:3). To whatever extent you intend to embrace theory, your interpretive passages are the place to draw it in and draw upon it. This is far preferable to a premature and abstract discussion of theory offered by way of introduction. Be forthright in how deeply you intend to delve. Theory ought to be *useful,* not simply for show. Roger Sanjek offers a practical lesson for drawing on theory quite different from making a parade of it. In describing how theory served as a resource in writing up an extended field study, he reports, "I searched for no more theory than I needed to organize and tell my story" (p. 3). If you are writing up research, theory should serve your purpose, not the other way around. When you can *make theory work for you,* use it. When theory is only *making work* for you, look for alternative ways to pull your account together and to explain what you have been up to.

Of course, *if* theory has guided your inquiry from the start, the reader should be informed from the start. But in observing students and colleagues at work over the years, I have more often seen theory *imposed,* in a too-obvious effort to rationalize data already collected, than I have seen data collection guided by a theory already well in hand. Field-oriented researchers tend to be greatly influenced (awed?) by theory. By the very nature of the way we approach things—flat-footed observers with feet of clay—we tend at most to be theory borrowers (or theory "poachers," as others sometimes see us)—not theory builders. Taking a model of theory-driven research derived from the so-called hard sciences doesn't serve anything but our already heightened sense of physics envy. Unless you think one must wear a white lab coat to be a careful observer, forget that model and keep your "theorizing" modest and relevant. Clifford Geertz observes in a brief new preface to a reissue of *The Interpretation of Cultures,*

This backward order of things—first you write and then you figure out what you are writing about—may seem odd, or even perverse, but it is, I think, at least most of the time, standard procedure in cultural anthropology. [2000:vi]

I'll hazard that it's standard procedure in *most* qualitative inquiry. Discovery is our forte.

Drawing theoretical implications is an important facet of the research process, and the advancement of theoretical knowledge is a reasonable expectation for the effort. But it should not be regarded as a *condition*. Advancing theoretical knowledge is not a step that every researcher is prepared, or *has been prepared,* to make. Take your work as far as you are able. Point the way for others if you are not prepared to take the theoretical leap yourself—especially if it begins to *feel* like a leap—rather than making a pretense at doing the "theory thing." If you have presented your descriptive account well and offered what you can by way of analysis (and interpretation), you have fulfilled the obligation of making your research accessible. Recognize that some scholars prefer to have us doing the basic descriptive work, freeing others who are more theoretically inclined to do what they do best. The way we continue beating up on the work of our predecessors should remind you that no one ever quite pulls off the whole thing. My hunch is that if you are drawn to qualitative inquiry, you are not among the theory-compulsive.

If you have the choice (that is, if you are not directed otherwise), consider integrating theory, or introducing your *concerns about* theory, into your account at the place where such concerns actually entered your thinking, rather than feeling obligated to slip theory in at the beginning as though it prompted or guided your research all along. One way to develop the "story" of your study is to discuss the point at which you (desperately?) longed for some theoretical or conceptual framework or some organizers or descriptors to help bring data together and make sense of them, at least to provide enough structure to guide your presentation. If one or two organizing concepts serve your purpose adequately, be content with that, extending a challenge to others of more theoretical sophistication—or daring— to ponder how *concepts* adequate for your purposes may fit into some broader theoretical scheme.

Theory should facilitate the inquiry process. It is not intended to bring quick and satisfying closure to an account. If you describe how and why you set out in search of theory (if, indeed, you really did), discuss what it was you felt you needed or hoped you would find. Your work will necessarily be unfinished but your reporting can be satisfyingly honest. Recognize in the conclusion of your writing that your search for theoretical insight may have only begun and is likely to persevere over a professional lifetime. That is why the problems that engage you need to be not only of genuine significance but of compelling personal interest as well.

The *search* for theory, like a cogent review of the literature, offers another way to link up with the prior work of others and a shorthand way to convey the gist of our interests and our inquiries. This "searching" stage is where one's dissertation committee, one's faculty colleagues, even anonymous reviewers, can—but seldom do—render invaluable service. Rather than belittle the efforts of novice researchers who thrash about trying desperately to hook up with theory, those more experienced in inquiry of this kind can—and should—suggest possible leads and links.[4] We all run the risk of getting tunnel vision when writing up our own research, failing to see the broader implications or remaining unaware of relevant work that might provide a fresh or clearer perspective.

This potential, and potentially invaluable, contribution that others can make needs to be given in the spirit of something *offered* rather than *issued*. With a gentle nudge, most of us can deepen or extend an analysis, yet a comparable and well-intended nudge toward theory can prove threatening. Doctoral students often reach this "Where's your theory?" point in writing their dissertations, pressed for time and feeling they have gone about as far as they can—or dare—go in theorizing their studies.[5] Potentially that presents a great teaching moment, provided help is proffered in a truly helpful way. When well-intended suggestions fail to take root, it seems preferable to leave fledgling researchers' accounts where they are rather than stepping in to wrest control from them. Wresting control may Save the Day for Science, but it comes at the possible cost of stopping beginning researchers cold in their tracks. Far preferable, it seems, for a student to submit an undertheorized study that is entirely his or her own than to feel that in the final moments, a work has literally been yanked (torn?) away, to be placed on a theoretical plane that the student is not yet able to attain.

Personal reflection: The satisfactory closure that my own dissertation committee was probably expecting, or hoping for, in 1964 did eventually get written. I appreciate that committee members were satisfied, if perhaps not wildly elated, with the essentially descriptive account that I presented. If they wondered among themselves whether I might be pushed to take things a bit further, they were kind enough not to insist. The consequence of such gentle treatment has been that whenever well-meaning colleagues or reviewers have pressed for stronger interpretations or theoretical implications in my work, I have continued to interpret their efforts as nudges rather than shoves. The studies have remained my own. If they strike some readers as undertheorized, my descriptive accounts have been adequate at least to allow others to do their own theorizing.

In contrast to my experience, I am haunted by the words of a student of a colleague who told me, years after the fact, that she never bothered to have her personal copy of the dissertation bound. "Why should I?" she queried. "Those weren't my words, they were my advisor's."[6] Such intrusiveness is most likely to be exhibited in theoretical heavy-handedness when a novice researcher is shoved aside by a probably well-intentioned advisor who insists, "Here, let me take over. You don't seem to know what you are doing at this point." More recently, a former colleague serving on a dissertation committee confided privately that he simply did not have time to bring the student's study up to his own high theoretical standard. Sound familiar? An academic put-down, when the offer of a kindly reach-down would have been so much more instructive.

ROLES THAT THEORY CAN PLAY

When used in daily discourse, the word "theory" signals no magic powers. It awards no special status to an idea, any more than other everyday words like "premise," "hunch," or "conjecture." I have friends who tout a fanciful "theory" that Hollywood film stars always die in threes, a lightheaded notion that proves accurate time and time again—provided, of course, that one is flexible about the counting.

But the term "theory" can become pernicious when used by researchers in its exalted "capital T" sense. Theory becomes a kind of intellectual bludgeon in the hands of reviewers, editors, or mentors who, intentionally or not, often intimidate applicants, authors, and especially graduate students anticipating or proposing a research agenda. Instead of capturing the teachable moment to share some possibly relevant *theories* (again, note the plural, never a single theory but several competing "hypotheses" offered for consideration), students are either *assigned* a theory or informed that they will not be allowed to proceed until they have identified an acceptable one. The idea was nicely captured in the title of a journal article that appeared years ago: "To Get Ahead, Get a Theory."

What might become an opportunity for an informed dialogue about theory, and a concomitant review of the roles that theory can play, is presented instead as an obstacle, much the way that the traditional "review of the literature" has usually been treated. Little or no recognition is given to the inherent danger that, in proceeding theoretically, objective reporting is often sacrificed in the grim determination to find what one has been searching for. And in spite of the intellectual appeal, there is little reward in setting out to *disprove* theory. Disproof may make for good science, but it does not make good copy. Articles in professional journals overwhelmingly favor, and thus report, favorable results.

In my resolve to keep a place for theory and yet keep theory in its place, I have proposed a modest definition in service to these dual purposes: *Theory is a way of asking (inquiring) that is guided by a reasonable answer.* If you have a reasonable answer (solution, explanation, interpretation, etc.) for the question you are pursuing, you can proceed with that as your focus. Whether or not you call your approach "theoretical" is up to you, depending more on your own needs or the demands put upon you by others.

But if you do *not* have a reasonable answer, or you are trying to figure out just what a reasonable *question* might be, why can't you make theoretical clarification part of your search? That is best done during the summing up, when you ponder the questions your observations pose for you and the possible contribution theory could make to guide the pursuit of explanations. Here is the opportunity for linking your work with that of other researchers. After presenting your study and making such connections, suggest what you think needs to be done next. This is where you can discuss the kinds of *theories* implicated by your work. Not to deny the importance of the roles that theory can play, but to guide you gently in the direction of making links with the works and ideas of others, rather than insisting that you display a theoretical sophistication that few ever achieve.

Most theoretical agonizing seems better located toward the end of a descriptive study rather than at its beginning. But must there be any agonizing at all? Would anything be lost by *playing* with theories, in the same way we sometimes claim to *play* with ideas? Similarly, it has been suggested that we need not, indeed, should not, limit ourselves to a consideration of only one theory at a time. Johan Galtung makes this plea on behalf of what he calls *theoretical pluralism* (Galtung 1990:101). Should you regard theory as too lofty even to make an appearance in your work, can you be coaxed into an examination of the *concepts* you have employed, or your *ideas,* your *hunches,* your *notions,* your *speculations,* even your *best guesses?* More modestly yet, might you make an initial foray simply by ferreting out critical *assumptions* that have guided your research? Don't kid yourself that you are above making assumptions. As long as you can state a problem and argue that one can reasonably "research" it, you've made assumptions aplenty, including some whopping assumptions about the nature and efficacy of *research* itself.

Another role theory plays—and could play to a greater extent—addresses a nagging shortcoming in qualitative study: our individual and collective failure to make our efforts cumulative. Every study tends to be one-of-a-kind, largely because of the fierce independence of most qualitative researchers and the limited scope of what any one individual can accomplish. A small step in this regard, in addition to a more generous spirit in recognizing the relevant work of others, might be for *each* of us to make better use of our own earlier studies in interpreting our later ones, to make our *individual* efforts cumulative over time, such as pursuing different aspects of a central issue or studying a common phenomenon from different perspectives.

For those of us already engaged in these efforts, a more self-conscious effort to grapple with theory in our *own* work, including considerations of how (or whether) theory might help overcome weaknesses in conceptualizing or linking new research, could serve a more instructive function than the repeated call we hear bemoaning the paucity of theory in the work of others. The challenge of making a greater effort to explore theoretical underpinnings need not, and should not, be placed so squarely on the shoulders of neophyte researchers entering our ranks. The responsibility belongs to those of us who have been at this work for years. It is an enduring problem, not something to be foisted on wave after wave of entry-level researchers.

We might also become more forgiving about our lack of theoretical sophistication in general. I am neither embarrassed nor apologetic about such a lack in my own work. I doubt that those with strong theoretical leanings find much of interest in my studies. I call my interpretations just that: "interpretations." I do not deny their implications for theory, nor do I deny that my data, like all data, are theory-laden; I subscribe to William James' notion (attributed in Agar 1996:75) that you can't even pick up rocks in a field without a theory. It is the term "theory" itself, and the mystical power attributed to it, that seem to get out of hand.

I like to quote Charles Darwin's advice about theory, written in 1863, because his own name is so clearly associated with theory and theorizing on a grand scale:

Let theory guide your observations, but till your reputation is well established be sparing in publishing theory. It makes persons doubt your observations. [quoted in Gruber 1981:183]

Darwin's caution deals with public claims-making, not with private theorizing. Indeed, the thoughts he expressed about the role of theorizing suggest that he felt it imperative to contextualize *all* observations within some larger purpose. Darwin referred to that larger purpose as The Grand Question. There seems little doubt in his mind just what that question should be:

The Grand Question, which every naturalist ought to have before him, when dissecting a whale, or classifying a mite, a fungus, or an infusorian, is, "What are the laws of life." [quoted in Barrett et al. 1987:228]

I presume Darwin did not badger his associates about theory, and I have not badgered mine. I am not theory-compulsive. More often, I have discouraged rather than encouraged students tempted to make too grand a leap from their modest observations to "theory," even to so-called theories of the middle range. I want my work to stand on its descriptive adequacy, theirs to do the same.

My claims-making is based largely on descriptive data gathered through fieldwork and organized around basic concepts from the social sciences. Individually and collectively, neither the techniques nor the concepts I employ are all that sophisticated. Wedded as I am to "culture" as a major orienting concept—a consequence of my anthropological heritage—I recognize the concern of colleagues who hint that the culture concept "points with an elbow" (i.e., is way too general). Perhaps concepts that point with an elbow offer direction enough, and theory enough, for anyone inclined to be cautious about, tentative toward, or dubious of theory (e.g., Paul Feyerabend's *Against Method* [1988]). I find solace in the words of those who soft-pedal the role of theory. Anthropologist George Peter Murdock expressed a view in his Huxley Memorial Lecture in 1971 that I continue to read as a caution for all field-based research:

> The quality of ethnographic description . . . seems to me, on the basis of my exposure to the literature, to depend remarkably little on the specific theoretical orientation of the observer. [1971:18]

What the quality of that description does depend on, Murdock continued,

> is not so much the theoretical orientation of the fieldworker, for this can probably produce as many blind spots as genuine insights, but rather such qualities as intellectual curiosity, a real interest in the people studied, sensitivity, industry, and objectivity. [p. 18]

The gist of Murdock's remarks was not to decry theory but to point to the disparity between observation and explanation for human behavior. Whether or not you concur with the set of attributes he identified for researchers (e.g., that a "real interest in the people studied" is a basis for assessing research acumen), the caution is that theory can get in the way as well as lead the way. Its roles and contributions to qualitative inquiry need to be weighed and examined, not exalted.[7]

LINKING UP THROUGH METHOD

If the role of theory tends to be underplayed in writing up qualitative research, the role and importance of method are more often overplayed, especially when method is used in reference to one's strategies for data gathering.

Fully explicated, method or "methodology" encompasses more than technique, far more importantly including procedures for data analysis. When method is taken in this broad sense as a way of doing something in accordance with a definite plan, then of course the term applies to the work of qualitative research, just as it applies to other activities that follow a consistent approach: acting method, birthing method, cooking method, organic gardening method, teaching method. There are books on all kinds of methods, and all kinds of books devoted to qualitative methods.[8] But when qualitative researchers address method as a topic to be "covered" in reporting their research, they tend to dwell too narrowly, too exhaustively, and sometimes too defensively on how they conducted their fieldwork and collected their data. It is that narrow sense of the term, method as technique, that I examine here.

I hardly take delight in repeating the observation that "no one who followed the scientific method ever found anything interesting" and then proceeding to laud qualitative research for its rigorous, systematic, hypothesis-testing procedures; its attention to measurement and quantification; its tight research designs; its concern for replicability—in short, with certain procedures that are not certain, not necessarily practical or appropriate, sometimes not even possible in the everyday settings where it is conducted. Paradoxically, those assumed to be wedded to the scientific method, our so-called hard scientists, are advantaged by what is essentially a self-perpetuating myth about how "sciencing" is done. The myth frees them to pursue their inquiries in ways that sometimes bear strong resemblance to our own. I remember a brief conversation with a seatmate on a transcontinental flight who said that he was a physicist whose specialty was the study of the ozone layer. In awe, I asked how one would initiate research on such a topic. I found his answer remarkably comforting: "First off, you need some observational data."

All research is based on observational data, an observation that is itself overlooked by those who insist that the difference between qualitative and quantitative approaches is that quantitative research is more objective. Pitting the two in opposition does a great disservice by detracting from the contribution to be made by each, including what each can contribute to the other. Most qualitative researchers would benefit by paying closer attention to counting and measuring whatever warrants being counted and measured; most quantifiers could "lighten up" to reveal how highly personal aspects about themselves strongly influence their professional practice. We all number our pages. We all make hopelessly subjective decisions in selecting the *topics* we research, regardless of how systematically some researchers proceed beyond that.

RESEARCH "TECHNIQUES" IN QUALITATIVE INQUIRY

A word of caution is in order for qualitative researchers tempted to lean too heavily on the sanctity of method, and especially to fieldwork techniques, to validate their research or to confer status. A critical "insider's appraisal" of the fieldwork techniques from which we derive our observations is appropriate here. That is the third and final kind of "linking" I examine. As with the previous two, I suggest you make less rather than more of this link, although the rationale is different.

When it comes to method, the links we can make to the work of others are neither powerful nor persuasive. Method is not the forte of qualitative research. Let me provide an overview to emphasize the rather ordinary, everyday approaches we employ. My intent is to dissuade you from a temptation to do likewise: You cannot build or strengthen your case by virtue of method alone. You are not obliged to review and defend the whole qualitative movement before you proceed with the particulars of your case. When it comes to careful explication of precisely how *you* went about *your* research, that is a different matter.

Prior to the past three or four decades, not much had been written about field methods. As best I recall, the phrase "qualitative research" was rarely (never?) heard in the 1960s. Of what had been written earlier, outside their respective academic disciplines, the same few references and the same few illustrative studies were cited almost to the exclusion of all others. Malinowski's introductory chapter in *Argonauts of the Western Pacific* (1922), "The Subject, Method and Scope of This Inquiry," was especially popular among those going far afield or searching out the exotic. William F. Whyte's *Street Corner Society* (1943) was an oft-cited text and model for anyone studying nearer home, particularly after Whyte added a 79-page appendix to a second edition (1955) that offered "a rather personal account of how the researcher lived during the period of study" (p. 279).

There were other works to be cited or consulted, a growing shelf of ethnographic classics for students of anthropology, augmented by studies in the "Chicago school" vein (see Deegan 2001) for sociologists. These constituted a manageable number of landmark studies whose titles were familiar even if their contents were not. Outside the fields of anthropology and qualitative sociology, however, they were generally regarded as exceptions to the rule of what constituted "real" or "rigorous" (i.e., experimentally controlled, and thus verifiable) research.

Today, a different circumstance prevails. Fieldwork "approaches" have been wrested from the disciplines that introduced and nurtured them. The techniques that characterize field studies are widely known and practiced. Darwin's caution about dwelling excessively on *theoretical* prowess might be paraphrased to provide comparable advice for qualitative researchers intending to rely on "method" to validate their claims: Let method guide your observations, but until your reputation is well established, be sparing in publishing about it.

How to Represent Qualitative Strategies Graphically

My enthusiastic embrace of qualitative research has been confirmed through years of doing it. My take on such practices has been influenced by training in cultural anthropology and applied, for the most part, in the broad arena of educational research. In addition to my research, I taught "qualitative methods" for years. For instructional purposes, I sought a way to represent qualitative approaches graphically. It was easy to summarize the approaches through tables that were convenient for class handouts or overhead projection, but a rigid columns-and-rows format gave the very opposite of the message of interconnectedness I wanted to convey. In what I understood to be the spirit and particular strength of qualitative inquiry, I sought ways to organize and present qualitative approaches as a seamless whole rather than through checklists of techniques.

Eventually I came up with the idea of representing the major approaches in a circle graph or pie chart. Circle graphs are ordinarily used to illustrate relative *proportion,* but it was the relative *similarities* and *inner-connectedness* among approaches that I wanted to emphasize. According to the way I placed the various approaches on the chart, the diagram allowed me to put in close proximity approaches that shared the most in common. That is the diagram I developed for the first edition of this monograph. (The original diagram is reprinted as Figure 5.1 and is discussed in the next chapter.)

I wanted to emphasize how it was not only the everyday nature of the data qualitative researchers dealt with, but an equally "everyday" set of techniques they employed for gathering those data. I felt that the data-gathering techniques could be subsumed adequately under three major headings. (That does not mean that I "found" three—there could be any number; instead, it reflects my preference for and tendency to *categorize* in sets of three.[9]) I assigned the labels *experiencing, enquiring,* and *examining* to the three categories. Those labels were intended to emphasize what the fieldworker actually *does* while engaging in them. You may recognize the three by their better-known labels: **participant observation, interviewing, and archival research.**

Conceptualizing Qualitative Approaches as a "Tree"

I began thinking about qualitative research as comparable to a researcher climbing into a qualitative research "tree" to get a view of what is going on. The picture forming in my mind was that of a giant broadleaf tree similar to the great oaks or maples I see from the windows of my study. Qualitative researchers avail themselves of the perspective they can get from whatever position they take in the tree. Each researcher makes a conscious choice as to where to get the best view for the information desired, and everyone realizes that it is impossible to be everywhere in the tree at once, although there are positions from which the view is said to be more holistic and complete.

I visualized the major research activities as roots penetrating deep into the events of everyday life through the three ways of gathering data: examining, enquiring, and experiencing. Firmly anchored in, and drawing nourishment through, these three roots is a solid "trunk" that represents the dominant core activity of qualitative research: participant observation. Emanating from the trunk are sturdy limbs that represent major variations of qualitative research. These limbs require varying degrees of personal involvement of the researcher at interviewing or observing. Each of these branches in turn has smaller branches that collectively comprise every possible strategy available from that branch. For example, one major branch leads to **archival sources** for its data and is dedicated to the **examination** of materials made by others (searching documents, studying artifacts). Another branch is devoted to **observation** (as with human ethology and all other types of non-participant observer study). Another branch focuses on **interviewing** (biography, journalism, narrative, oral history).

There is obviously some researcher involvement in each of these, but the degree of participation becomes more critical as one gets farther up into the tree where the trunk bifurcates into two branches, one extending into the more anthropological tradition of **ethnography,** the other encompassing the **field study** of the sociologist. Participant observation remains a central element in both approaches, although each has developed more specialized branches that take one beyond the limits of the ethnographic experience itself, as with the micro-ethnography of the anthropologist or the symbolic interactionism or ethnomethodology of the sociologist.

Researchers who seek the broadest perspective do not venture onto branches that depend on only one major strategy (e.g., a study conducted *essentially* through interviewing or archival research). They prefer a coign of vantage[10] from which they can combine strategies as needed. Their studies employ a wide array of techniques subsumed by the broad label **participant observation**. Anthropologists use their location to gain their broad ethnographic perspective or follow along a more highly specialized secondary branch, such as the *community study* or the *anthropological life history*. Or, their strategy may be a contemporary offshoot such as *critical* or *feminist ethnography*. Along a parallel sociological branch, one may secure a position in the sociologically oriented *field study* that became known as *"Chicago school" sociology* or look for a more specialized branch such as *phenomenology*, or *symbolic interactionism* (see Figure 4.1).

I present the "tree" sketch here to suggest (and attempt to illustrate) that there is little point in trying to provide a grand overview of qualitative research when any particular study can draw only selectively among such a wide variety of techniques and approaches. Broad overviews are properly the subject of entire books devoted to the topic. The label *participant observation* contributes to the confusion because it is sometimes used as the cover term to refer to *all* qualitative approaches, but can also refer to a single strategy among them (in which case the project is identified as a *participant observer study*). Thus, it is essential to provide detail as to exactly *how* participant observation, in this all-inclusive sense, is played out in any *particular* piece of research. The label itself is too encompassing.

Venturing up or out from the security offered by the dominant core activity of participant observation requires conceptual and methodological know-how, just as each approach (each major "branch") offers particular advantages and its unique perspective. One can also go "out on a limb" by claiming to achieve some special perspective, yet failing to realize either the limitations or the potential to be gained from that perspective. Beyond a generic "participant observation" approach, there are particular ways of conceptualizing associated with particular disciplines, labels, schools of thought, or problem orientations. Ethnography, human ethology, oral history, symbolic interactionism—each of these offers a particular way of seeing and a particular perspective on what is observed and how it is presented. Yet, as Kenneth Burke noted years ago, "A way

Figure 4.1   Portraying Qualitative Research Strategies Graphically

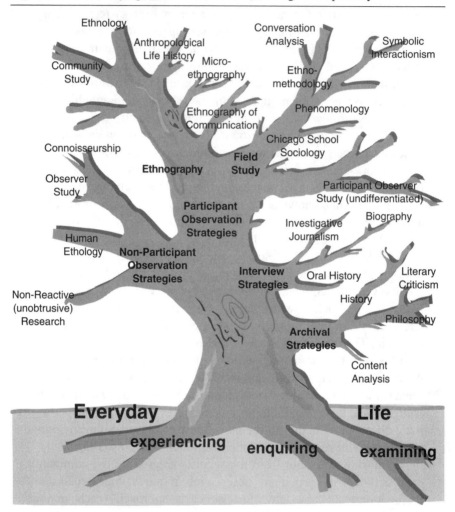

of seeing is also a way of not seeing" (Burke 1935:70). Whether in a figurative tree or a literal one, you cannot be everywhere at once or afford yourself of every possible viewpoint at the same moment.

Some unanticipated things happened when I transformed that original circle graph into the tree diagram. For one, the visualization helped me realize how participant observation, broadly conceived, serves as the core activity for *all* qualitative work, rather than simply as one *facet* of it, the way I had represented it in the original diagram. That is why participant observation doubles as a synonym for

fieldwork, for ethnography, for virtually any approach that is "qualitative." Participant observation *is* at the heart, and thus *is* the heart-
wood, of all qualitative inquiry, its substantive core.

I was also surprised to discover that I had trouble finding a suitable place for *case study* on the tree. My problem was not that "case
study" did not fit *anywhere* but that it seemed to fit *everywhere.* I
pondered that perhaps case study is better suited if it, like all other
ways of organizing data, is regarded as a genre for *reporting* than as
a strategy for *conducting* research.

I recognize that some scholars consider the case study to be an
eclectic but nonetheless identifiable *method* (e.g., Merriam 1998;
Stake 1995; Yin 1994). If you intend to present your study in the
form of a case, or a narrative, be sure to provide adequate detail
about the specific fieldwork techniques you employ, rather than hope
that by itself your label provides enough detail about how you proceeded. Like any of the generic strategies mentioned here, you could
write an article or book about it, but for any given study, the label is
woefully inadequate as an explanation of how you proceeded.

I like analogies. I have employed the tree analogy as a way to
explain, and the diagram as a way to illustrate, how participant
observation constitutes the core of qualitative research. The visual
also helps to emphasize connectedness among the approaches: what
they share in common and how they are differentiated. Not only
must choices be made among strategies, but commitments must be
met in making them. For example, there is a vast difference between
borrowing one or two of the fieldwork techniques that ethnographers
(and other qualitative researchers) use to gather data and getting "out
on a limb" by claiming to be "doing ethnography" solely on the basis
of technique. A study *influenced* by an ethnographic approach, or by
phenomenology, and so on, is not the same as a study *well grounded*
in these approaches. Such a study is best represented as "in the manner of," rather than laying claim to demonstrate all of the nuances of
seasoned researchers fully conversant with that tradition.

I hoped that others would find my tree diagram and analogy
engaging and would feel free to add or rearrange the branches or
develop sections of it in greater detail.[11] I assumed that the portion
of the tree representing a researcher's particular interest—as with the
anthropological "branch" in my case—would be portrayed with relatively greater attention to detail. Areas of less interest would be
treated cursorily, not out of neglect but to call attention to some particular subsection rather than to the tree itself.[12] That should be true

of the literature on method that you include in your review. Your readers do not need the whole story of who may have planted the tree (it always seems to get back to Herodotus) or how the tree has evolved and matured. Your readers need to be assured that you are secure in the position from which you draw your perspective and that you are also aware of the limits on what you can and cannot see. Your selection of a research stance should be a reasonable and reasoned one, well suited to your purpose and your talents.

Your Methods; Methodology

I recommend that you not devote undue attention to a general review of method. If a dissertation committee wants assurance of your command of that literature, here is another aspect that you might develop in your initial research *proposal,* subsequently to be employed selectively on a when-and-as-needed basis. By identifying participant observation as the *core* research activity in qualitative inquiry, I have underscored not only the everyday nature of what we study but the everyday nature of the way we go about collecting data. It is impossible to shroud in mystery an approach that can be encapsulated by the term "participant observation" or an alternate label attributed to anthropologist Renato Rosaldo, "deep hanging out" (in Clifford 1997:188).

Fieldwork techniques alone are not sufficient to allow us to make vigorous claims about what we have done. Employing them in the course of an inquiry does not require one to dwell excessively on who has pioneered them or who has employed them elsewhere. Neither "being there" nor "intimate, long-term acquaintance" is sufficient to guarantee the accuracy or completeness of what we have to report. There is little point in trying to make a big deal of them (see also HFW 2008a:xi).

Qualitative inquiry consists of more than method, and method is more than fieldwork techniques. The more you dwell on the latter, the more you draw attention from your substantive findings. Don't try to convince your audience of the validity of observations based on the power of the fieldwork approach. Satisfy readers with sufficient detail about *how you obtained the data you actually used.* You are the best source of information about the confirmability of what you have reported. If you level with your reader on that score, you will have fulfilled the obligation of careful reporting. But the potential of your contribution will be greatly enhanced if you provide adequate detail about how you proceeded with your *analysis*. The unique combination of your field setting, and you immersed in it at some particular moment, is not likely to be replicated, but discussing how you analyzed your data might be a great help to other researchers with comparable field notes, experiences, and data sets of their own to analyze.

If you have lots to say about doing fieldwork—plenty of good advice or cautions to pass along—write up such material separately, perhaps as an appendix, maybe a separate article or monograph. Should an inquiry prove disappointing in its substantive contribution but have an important contribution to make to method, be sure that your changed focus is communicated clearly by revising your statement of purpose. Don't hesitate to provide rich examples from your fieldwork experience to illustrate your points, but don't try to satisfy diverse audiences by trying to achieve a balance between substantive issues *and* method.

Should you want to give particular emphasis either to *method* (fieldwork techniques, analytical procedures) or to *methodology,* write those pieces separately and with a method-oriented audience in mind.[13] If method is, or becomes, the focus of your account, note the distinction that can, and should, be made between *method* and *methodology.* Methodology refers to underlying principles of inquiry rather than to specific techniques. I don't know where we picked up the habit of referring to everything we do as "methodology." I do think I know *why* we do it: It makes our humble techniques seem more highfalutin'.

METHOD OR THEORY AS NARRATIVE

The thought may have occurred to you that the events of your research as they unfolded, from how you first became involved in the topic to how you proceeded and what you learned, might be an appropriate vehicle for narrating your account. With only the caution that you keep the narrative focused on the research, rather than on the researcher, that might be an excellent approach. It allows you to introduce vital supporting information as the account evolves, rather than to begin by devoting traditional chapters to broad discussions of the literature, relevant theory, and "method." I note and commend this approach; I would not want to leave the impression that method is to be slighted. The operative word in making any of the links discussed here is *relevance.* Resist the temptation to fill pages with topics too broadly defined to move the account forward.

In rare cases, at least among novice researchers, the pursuit of *theory* might similarly provide a narrative thread for weaving the account together. More seasoned hands might be able to weave a spell-binding tale of theory-driven research, especially if the account reveals the risks of conducting research when you know in advance what you are looking for. Personal experience offers yet another way to develop your story, perhaps related as a problem originally encountered in your personal life that later is recast in terms of professional career. Do not rule out any approach that helps you tell the story you have to tell and to maintain your focus as the account becomes more complex.

CODA

This concludes a discussion that has reviewed how to get and keep you going and what to include in the body of your text. From here, we move to questions of tightening up, finishing up, and getting published. In this chapter, I turned from the mechanics of writing to questions of "what else" to include that supports what you have written. I have addressed some traditions of academic writing, traditions that have their origins in dissertation writing but tend to linger because "we've always done it that way."

Although I have cautioned against getting on a soapbox in writing up *your* study, that has not prevented me from getting on a soapbox of my own in this chapter. I intend these comments to be provocative. If you have been unwilling to "buck" tradition in writing up your first study, I encourage you to think about alternative ways to present material when you are on your own. Look for ways to encourage others, your students especially, to do the same.

I have proposed that your Chapter Two should be whatever you think Chapter Two should be. If you are going to blow away Chapter One with the usual academic throat-clearing introductory comments (another hard-to-break habit), by the time you get to Chapter Two it is time to get to cases. Whatever compels *you* to provide an extensive review of the relevant literatures, your reader would probably prefer that you get on with what you have to report. To that end, I suggest that after only the briefest description of the problem and scene, you turn immediately to the details of your account. Introduce additional contextual material on a when-and-as-needed basis to make sense of what you have presented and to situate your work within a broader context.

The same with theory and method: when-and-as-needed, with a view to explicating, rather than interrupting. Regard theory as "user friendly," an invaluable resource if you know how to use it. If you aren't quite there yet, share your musings about how the right theory at the right time might have helped. That seems preferable to imposing some tangential theory that lends form rather than function. Nobody *insists* that dissertations must read like dissertations. The awkward way that theory gets injected into most of them can be part of the problem. If practical solutions, or broad concepts, or the drawing of analogies, have been adequate for organizing and presenting your data, talk about them. Don't be tempted to scratch about for some lofty theoretical notion that may obscure the importance of the observations and insights you have to offer.

In the past two decades, qualitative methods—which in many instances would be more accurately portrayed as fieldwork techniques—have come to be widely known and accepted. There is no need for each researcher to discover and defend them anew, nor a need to provide an exhaustive review of the literature about such standard procedures as participant observation or interviewing. Instead of having to describe and defend qualitative approaches, as we once felt obligated to do, it has become difficult to say anything new or startling about them. Neophyte researchers who have recently experienced these approaches firsthand need to appreciate that their audiences are unlikely to share a comparable sense of excitement about hearing them reviewed one more time. Say less rather than more in defense of qualitative research in general.

Do not rely on method to validate your work. We are more systematic, more "methodical," than our critics give us credit for, but we could spin our wheels forever trying to convince them. Such discussions find us more often on the defensive than on target about what we make of what we have observed. Method is not what we are about, at least in the sense of relying on technique to affirm that we've gotten it right.

Especially in your post-dissertation writing, keep other researchers in mind as an audience—and perhaps as *the* audience—most likely to be deeply interested in your work. They are not interested in a display of all you know, but they do expect you to demonstrate a command of the *relevant* literature of your field.

They will be interested in how you grapple with theory, and especially with your ability to link theory with your approach. Because your data consist essentially of rather everyday stuff collected in rather everyday ways, any insight you have gained about organizing and analyzing data will be especially welcome. Truth be known, the real "work" of qualitative research lies in mindwork, not fieldwork, as others have pointed out (see, for example, Agar 1996:51).

Summing Up: Major Linking Activities

• If absolutely required by a dissertation committee, editor, or publisher to address such topics as a stand-apart literature review or a separate treatise on theory or method, do as you must. But do not simply *assume* that these topics have to be dealt with in a particular place or manner. Be selective in everything you include by way of review, and do not let your review of what others have done overshadow or preempt what you have to report.

• Do not overdo the lit review. Draw attention to closely related studies. Save broadside review efforts for separate writing assignments. If you are writing a thesis and the comprehensiveness of the literature review required by your thesis committee seems to far exceed the scope of your research topic, see if you can negotiate to prepare such a review separately, so that anything you include in the thesis itself has immediate relevance.

• Do not cower before theory. If theory has neither explicitly guided the research nor been of help in the analysis of data, discuss what you hoped theory *could do* for you and the likely form(s) that help might take, rather than trump up some nebulous theoretical links that serve only as window dressing. But do track the origins in your thinking about the problem you are investigating, its significance, its complexity.

• Do not belabor the broad topic of "method," and do not attempt to review or defend the entire qualitative movement. Restrict your detailed explanation of fieldwork techniques to how you obtained the data you used, not how everyone who pursues a qualitative approach goes about getting theirs. Same with analysis: What did *you* do that made your data "usable"?

NOTES

1. The special interest some academic advisers have in a comprehensive lit review may reveal their personal struggles to remain current in their field. They may use and rely on their advanced students to scour the research literature that they feel they cannot keep up with. You might consider "keeping-the-faculty-up-to-date" a hidden cost of preparing a dissertation, but it does afford an opportunity to contribute—if only indirectly—to academic scholarship.

2. For a discussion on how to make qualitative work appear less boring, see Caulley (2008). His antidote for boring writing is to use techniques from *creative nonfiction,* writing nonfiction but using techniques borrowed from fiction (see also Cheney 2001).

3. For more on conducting a literature review, see Fink (2005); Pan (2008).

4. Is there any reason why making the theory link could not become a more interactive (and more collegial) element in dissertations, with faculty input publicly acknowledged by identifying not only the theoretical insights proposed but also identifying by name the faculty member who suggests them?

5. For a discussion of how theory can be incorporated into descriptive studies, accompanied by illustrative examples from their recently completed dissertations by several then-new PhDs in educational research, see Flinders and Mills, eds., *Theory and Concepts in Qualitative Research* (1993).

6. With electronic storage, there may be less need today for having one's personal copy of a dissertation "bound," but the significance of these remarks was that the student had no pride in the final product because she did not consider the work to be her own.

7. See, for example, the discussion of the influence of theory on anthropological explanations in Layton (1997).

8. For example, Russ Bernard's *Social Research Methods* addresses both qualitative and quantitative approaches (Bernard 2000), or the comprehensive *SAGE Handbook of Qualitative Research,* edited by Norman Denzin and Yvonna Lincoln, now in a third edition (2005) containing 45 separately authored chapters. See also Creswell (2007).

9. Catch yourself if you fall habit to this common reporting error, treating the number of categories we impose on phenomena (such as the three categories suggested here) as a quality inherent in the data themselves.

10. Coign of vantage refers to "a good position for observation, judgment, action, criticism" (Random House unabridged 2nd ed.), which is exactly the way it is used here. I encountered the phrase in the first dissertation I read at Stanford (by a fellow named Thomas, as I recall), and I resolved to use it someday myself, if possible, to show that I had achieved

the same level of scholarship as someone who had actually *completed* a dissertation. Now, some 50 years later, my mission has been accomplished.

11. Instead, my efforts to illustrate these links seemed to have gotten out of hand when an overzealous illustrator rendered the tree in full leaf. The editors who solicited the chapter in which the tree originally appeared, joked about the figure, which they nicknamed—respectfully, they insisted— "Harry's burning bush" (LeCompte, Millroy and Preissle 1992:xxiii).

12. Developing the tree diagram would lend itself superbly to a poster session, for the tree is a working draft, not a finished product; it still needs refining and would benefit from the input of others. I have not intentionally left out any strategy. I do emphasize the importance of the three major roots that represent how we get data.

13. Several qualitatively oriented journals include articles about research methods. The journal *Field Methods,* formerly *Cultural Anthropology Methods,* is devoted exclusively to articles about the methods used by field-workers in the social and behavioral sciences and humanities.

CHAPTER FIVE

Tightening Up

. . . and now I am trying to do it again to say everything about everything . . .

—*Gertrude Stein,* Everybody's Autobiography, *p. 80*

Some of the best advice I've ever found for writers happened instead to be included with the directions for assembling a new wheelbarrow: "Make sure all parts are properly in place before tightening."

One can press the analogy. Fieldwork and organizing one's data might be likened to collecting and identifying the "parts" of a wheelbarrow. Once you have gathered all the parts, you need a basis for sorting them and a workable sequence for assembling them. Think how you will proceed. Do you have everything you will need? Conversely, do you need everything you have? Remember, you're only supposed to be tightening that wheelbarrow, not filling it!

Whether writing up your research entails for you a project report, a journal article, a thesis, a technically oriented monograph, or a book, my guess is that as the material takes shape, you will worry that the descriptive account is too long, the interpretation or analysis lacks the power you hoped to achieve, and you cannot figure out what to say in conclusion. Dismaying as these problems may appear, you are nonetheless making headway if they are yours. Think how much better off you are than a researcher who discovers that the data are thin, the analysis and conclusions unwarranted, the basic research question misguided.

No amount of editing can transform an inadequate data base into a solid piece of research, although candor on the part of the researcher may preserve ideas, insights, and questions that have merit independent of fizzled fieldwork. But don't fool yourself that some fancy combination of boldface type, shadow letters, underlined words, or whatever format or lettering tricks you can do with word processing or color printing can improve anything except the "looks" of a manuscript. (Fancy title pages on term papers always aroused my suspicion.) Qualitative studies are judged by how they read, not how they strike the eye.

Fieldwork may be the dramatic element in qualitative work, but the real test lies in the way everything is assembled in the final product. That is why "method" is more than collecting data, and more than simply reporting data one has collected. Writing is not the only thing involved, but it is the focus here. Writing is integral to qualitative inquiry, not adjunct. Some researchers achieve brilliant results through seemingly effortless prowess with prose, but what most of us accomplish is achieved through sustained effort at editing and revision. Those are the processes discussed next. As for emulating Gertrude Stein and trying to say everything about everything, rest assured it is not the way to go about writing up qualitative research.

DESCRIPTIVE ADEQUACY

"When in doubt, leave it out," the guidebooks advise the traveler packing for an extended trip. Good advice for qualitative researchers, too—although when we are unsure about how much to pack into our accounts, we're more likely to do just the opposite. How much description is enough to earn the accolade "thick description"? How much context is enough to make a study "contextual"? To avoid being shallow, how deeply must we delve to present a case "in depth"? Yet if luminaries like Malinowski or Margaret Mead can be faulted for "haphazard descriptiveness" (noted in Marcus and Fischer 1986:56), how can we be sure that our own descriptive efforts attain some higher order?

Faced with the dilemma of having more to pack than a suitcase can possibly hold, the seasoned traveler has three possibilities:

rearrange to get more in, remove all non-essentials, or find a larger suitcase. Qualitative researchers face comparable alternatives. Like learning to pack small items inside bigger ones, there are ways to pack more into a manuscript without increasing its length.

"Tightening," of course, implies that the end product will be more compact, although experience suggests otherwise. Unless revision is undertaken specifically to reduce manuscript length, my deletions are usually matched by seemingly minor changes and additions that leave total manuscript length about the same—or a tad longer. If clever repacking is not sufficient, some items simply have to be left out. As to the third possibility, travelers and seasoned researchers alike are aware that large containers (like bigger volumes, or two-volume sets) are unwieldy, often require special handling that adds extra costs, and may be prohibited by regulation.

Under the headings that follow, "Packing More In" and "Tossing More Out," I offer suggestions of a mechanical bent for helping authors comply with space limitations. The underlying concern is not mechanical, however; it is an issue of focus. That is why I keep returning to the importance of the problem statement, "The purpose of this study is . . . ," and reiterating that the problem statement itself must remain under continuous review.

Packing More In

Given lingering doubts about the criteria of inclusion for descriptive studies, I can offer another aphorism that has served to guide my writing and that I have frequently repeated to help others experiencing difficulties with organizing, writing, or editing: Do less, more thoroughly!

"Do less, more thoroughly" is my maxim, and the zoom lens on a camera provides an analogy for the principle in action. If you want to take in more of the picture, you must sacrifice closeness of detail; if you want more detail, you must sacrifice breadth. Michael Agar suggests what he calls the "funnel approach" to fieldwork: "The strategy is to *selectively* narrow the focus within a previously explored broad field" (Agar 1996:61; for another lens analogy, see Peacock 1986). Do you have the focus right? Keep in mind that the answer to that crucial question lies not within the research setting, and not within your choice of method; it is something you *impose* on the setting.

In spite of growing interest in the performance of text, qualitative research weds us to prose. But we are not limited only to words. Charts, diagrams, maps, tables, and photographs provide valuable supplements to printed text and help condense and expedite the presentation of supporting detail. I have noted my inclination to "think sections," "think chapters," or "think Table of Contents" from the moment I begin a study. That advice can be restated more universally, applicable to the presentation of quantitative and qualitative data alike. Miles and Huberman state succinctly: Think display (1994:11 *passim*).[1] Display formats provide alternatives for coping with two of our most critical tasks, data reduction and data analysis.

Charts and diagrams offer additional ways to give our thoughts "embodiment." They invite us to sort and to categorize data, to explore what-goes-with-what, to contemplate how seemingly discrete data may be linked in previously unrecognized ways. Where most of us are constrained by our regimented vision of prose, researchers who think spatially work through their charts and diagrams in order literally to "see" their studies. From the outset, some qualitative researchers conceptualize their studies in charts and diagrams drawn on inexpensive newsprint spread across their walls or floors. Anyone who has been a presenter in a poster session has experienced the challenge to "think display."

In more conventional formats, tables and charts can also relay or summarize information that provides context for a study but are of interest to only a portion of one's readers. Similarly, maps are an expedient way to locate a region and community, sketch maps a convenient way to plot movement or show before-and-after comparisons, and pictures are still worth a thousand words . . . more or less, and with the caveat that they do not raise insurmountable issues with confidentiality or permissions.

Graphics also enhance the likelihood of capturing the attention of readers who "see" facts or visualize relationships in other ways. They keep us mindful of exploring alternative forms of representation and presentation by augmenting the always-potentially-tedious flow of words on a printed page.

Display has a function in data *analysis* as well. Charts and diagrams developed in rough form during preliminary efforts to organize data (as well as get rid of it) can help researchers tease out relationships and patterns spatially. Don't hesitate to explore alternative ways *and shapes* for displaying and summarizing data. But do keep them easy to follow, easy to understand, and purposeful rather than decorative.

By way of example, I reproduce here a diagram that I prepared for the first edition of this monograph (see Figure 5.1). The diagram was designed to provide a visual representation of the major approaches to on-site research. One purpose for organizing the material was to emphasize a variety of strategies, so that students new to qualitative research would not attach a catchy label like "ethnography" to their studies merely because they assumed that a single label served for them all. I wanted to convey a sense of the interrelatedness among approaches without implying a sense of hierarchy. A circle diagram provided a way to represent the approaches in a continuum ranging from closely related ones to seeming "opposites." It was a beginning of the effort that finally resulted in the more satisfactory tree diagram presented as Figure 4.1 in the previous chapter.

There are already more words in the previous paragraph than in the figure, which is another part of the message: the chart (almost) speaks for itself. That the chart speaks for itself happens to be both a fact and a precondition for material that appears separate from the text: Accompanying material must stand independently. Supplementary material must be adequately labeled so that it can be understood without having to consult the text, a function performed with captions and subheadings. The challenge is to ensure that the material is self-explanatory, not text-dependent. To determine whether your charts and tables stand by themselves, ask someone unfamiliar with the text to interpret your visual displays.

Captions and explanatory material accompanying tables, charts, and photographs deserve editorial review. And nothing should be considered for inclusion that is not of high quality (i.e., clean lines, sharp image). Materials also must be relevant to the purposes at hand, not used simply to break up space or create an impression. As revising and editing continue, so should the review of supporting materials. Charts or tables critical to early drafts may become superfluous as the writing proceeds.

Conversely, maps and diagrams may have to be simplified to be effective. Several maps may be necessary to accomplish competing purposes originally designed to be achieved by one (e.g., locating the region of study in one map, accompanied by a map of larger scale providing important local detail). Don't leave the reader

Figure 5.1   Qualitative/Descriptive Studies Organized by Research
Approach

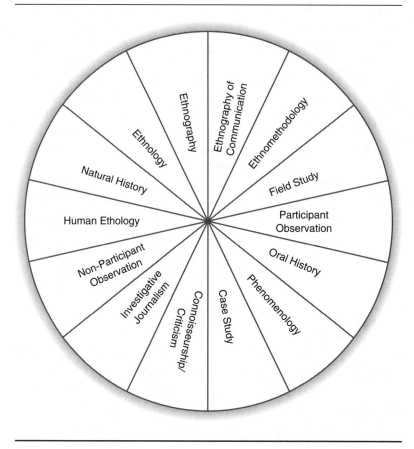

wondering why something was included that seems to bear little
relationship to the purposes of the study.

A word of caution: Charts and tables provide a ready trap for
authors susceptible to what Lewis Coser labeled "misplaced precision," the effort to compensate for theoretical weakness through
methodological strength (Coser 1975:692–693). Misplaced precision is not inherent in charts and tables themselves. It becomes
apparent when data are introduced that are distracting, or are incon-
sistent with the purposes of a study or with the level of detail provided elsewhere. (As a dyed-in-the-wool qualitative researcher, the

more charts and tables I find in a study, the more suspicious I become that the researcher may be trying to impress rather than to inform.)

In the case of my original circle diagram, however, it was its *lack* of complexity that concerned me. Models should simplify, but this one oversimplified by representing each of the 14 qualitative approaches as independent and equal. Initially, the diagram served my intended purpose, to present an overview, but the complexity of the relationship among qualitative approaches called for more. The ever-evolving tree diagram described previously was the result. My pie chart was only a beginning that led eventually to something more satisfactory.

When they summarize or illustrate important information, tables, charts, photos, and similar figures render valuable service. They also broaden the appeal for those who appreciate having data presented in non-textual form. But the presentation of data can itself become a preoccupation. Biographers and historians seem particularly susceptible to the temptations of data overload, sometimes seeming to include data for no apparent reason except that they were uncovered during the course of research. The most flagrant example of misplaced precision in qualitative reporting that I recall was the inclusion of a table of random numbers that appeared as an appendix in a monograph on social practices related to beer drinking among urban Africans. The intended message, I assume, was, "This is science." My reaction was, "This is ridiculous."

Another practice that can help keep manuscripts to a reasonable length is to provide only brief excerpts from interviews or field notes in the body of the text. When necessary, longer protocols can be included in an appendix or supplement, an alternative I discuss briefly in Chapter 6. True, shifting data from one place to another does not change the overall length of a manuscript—it relocates the problem rather than alleviating it. Editorially, however, it helps to lend emphasis and focus by drawing attention to critical elements rather than simply turning readers loose to forage for themselves. It helps guard against the temptation to let informants prattle on, as they may have done (and may have been encouraged to do) in the original interviews. Allowing informants to "go on and on" in the written account can be counterproductive, more likely to leave readers bored than beguiled.

At times, informants do need to be given their voice, and there are qualitative approaches such as the life history in which that voice may be the only one heard. In general, however, brief quotes are usually more effective than lengthy ones, especially when multiple speakers are addressing the same topic. The longer your quoted passages (whether from informants or from printed sources), the greater the need to ensure that your reader understands the point *you* are making. Sometimes, by first relegating longer sections of interview material or quoted text to footnotes or supplementary appendixes, you may realize that they can be eliminated altogether. Once separated from the text, it is easier to judge how vital a contribution they make.

Tossing More Out

When outside reviewers agree that some part or parts of a manuscript can be deleted, I hasten to follow their recommendation. In spite of general consensus about the need for cutting, my experience is that developmental reviewers will disagree among themselves as to precisely what to cut, what should stay. If cutting is both major and mandatory, I appreciate the authoritative voice and experience of an editor for suggestions about how to proceed.

I have discovered that invited readers of early drafts are more apt to identify "possible" cuts if I explain that my problem is no longer whether to cut, but where? To reassure them that I am serious about cutting, I also discovered that if I provide a draft on which I have already marked a few deletions of my own, reviewers are more likely to render the help I need than if I present them with a clean draft. Some reviewers are reluctant to make notations on clean copy, so the copy they receive has a few changes already marked.

I never ask anyone to review copy that is difficult to read or heavily marked on, nor would I ask anyone to read copy that is unpaginated or unproofed for spelling. I think it is rude to do so and totally unforgivable in a day when turning out clean, correct, and properly formatted copy is so easy to do.

Whatever the motivation for cutting, whether on the recommendation (or insistence) of others or an intuitive feeling of your own, you should always do the cutting yourself, along with any necessary rewriting, rather than delegate it. If you do not have access to critic readers willing to help you identify possible ways to cut, here are some places where you can begin a search of your own.

First, look for little diatribes where you may have gone off on a tangent. These become easier to identify after you gain some distance from your manuscript. You may suddenly realize that certain questions or issues were dropped rather than developed, or that you took advantage of an opportunity to get on a soapbox about a topic of perennial concern *to you* but not of vital interest to your readers. Colleagues can help spot such detours if you specifically ask, although they may be too polite to mention them if you do not. They are, after all, familiar with you and your ways.

Second, make a critical assessment of all points supported with multiple illustrations, multiple vignettes, or multiple quotations, with a sharp eye for repetition. Save the best and drop the rest. Summarize the general pattern you see, retaining an illustrative example or two. You may be attempting to preserve and portray important nuances that *you* recognize among closely related examples, but subtle differences are likely to go unrecognized, and therefore unappreciated, for readers lacking your firsthand experience. Most of us see and even "hear" our informants as we enter their words onto a manuscript. We forget that our readers cannot do that; to them, the words remain lifeless except for the voice we give them. The repetition of materials that *appear* virtually identical can be tedious.

Don't be hesitant about including an excess of illustrative material in your early drafts; you can winnow the material as you edit. Here we are talking about how to tighten up a working manuscript, not about what goes into the initial draft. It is easier to synthesize or delete from too many examples than to go back through notes searching for an illustration or example dropped too soon. Your choice of quotes and vignettes also may change as the text develops. Be sure to mark your excerpts (preferably where they appear in your manuscript) so that you can quickly locate original sources. Your coding system also can serve as a reminder and relay important information in encapsulated form, such as "Field notes, 11/8/84," or "M-23" [male, 23 years of age].

Third, examine carefully every beginning: the first sentence and paragraph of each chapter or section, the first section of every chapter, even the entire first chapter itself. In rereading my own drafts, I have discovered that my start-ups often prove little more than warm-ups. They helped me to get the momentum going but do nothing for readers likely to be cooled out by a slow start. You may be able to lop a bit off *every* beginning—unless, of course, you heeded my advice to plunge into your account with that key sentence, "The purpose of this study is. . . ." Several readers of early drafts asked why I hadn't begun this monograph that way, politely hinting that my first chapter, brief as it is, still seems to get off to a slow start.

Let me express one caution about beginning with the phrase "The purpose of this study is. . . ." By the fifth word into your text you will already have used the expression "this study." Reference to "this paper," "this dissertation," "this study," or "this book" must be among the most frequently repeated phrases in all academic writing. Search and destroy such phrases when possible. Give your readers credit for being able to remember what they are reading.

Finally, look for whole intact *sections*—even entire *chapters*— that might be dropped or relegated to a separate writing project. I have already proposed (in Chapter 4) that an extended discussion of method is a likely candidate for a separate paper. You may discover that huge chunks—paragraphs, sections, a whole chapter—can be deleted where smaller pieces cannot. Deleting big chunks may leave glaring gaps that you can point to as major topics intentionally passed over, to be dealt with at another time.

At the suggestion (I like to think of it as a suggestion) of my original editor, and in order to stay within space limitations imposed on all monographs originally published in the Qualitative Research Methods series, I deleted *two entire chapters* from the first edition of the monograph. At the time, deleting them seemed to me to leave a gigantic hole in the monograph. But no one seemed to notice, and each of the deleted chapters was subsequently developed into a separate article that I was able to publish elsewhere.

By contrast, fiddling with minor cuts may leave you feeling anxious that the account is becoming choppy and disjointed, perhaps even losing its integrity. Mitch Allen recalls advising me that another of my book drafts "needed to go on a diet." He recommended I try to cut three pages from each chapter. At the time, that seemed a reasonable suggestion, something akin to the traditional New Year's resolution about losing a few pounds. As Mitch has since reminded me, "I don't think you ever got there." Was he really aware that the manuscript at the time had 11 chapters? He wanted it to be $11 \times 3 = 33$ pages shorter! Isn't a little dieting better than none at all? Granted, pruning stimulates growth, as our gardening friends tell us. Editorial "pruning" also ought to be invigorating, for author and manuscript alike. But I always find it hard to do. And I always keep a copy of my earlier and longer version(s). Unlike the gardener, we do get a second chance should initial cutting prove misdirected or too severe.

How to fit qualitative-descriptive research into the prescribed limits of journal-length articles or monograph-length series, or how to compress two years of fieldwork and writing into a fifteen-minute time slot in a symposium, is vexing indeed. David Fetterman has reported a rare case where management objected to the *brevity* of an evaluation report he had drafted, but he notes that the objection was based on the belief that "a physically weightier document would be more useful to them to sell the program in the future" (1989:17). That may be the only occasion ever recorded when a qualitative researcher was asked to lengthen an account.

Faced repeatedly with the dilemma of having to compress qualitative reporting, both in my own writing and in trying to help students and colleagues with theirs, my resolution, advice, and philosophy is summed up in the idea of "doing less more thoroughly." A strategy for accomplishing this is to look for parts or instances or cases that can stand for the whole. *Synecdoche,* our literary colleagues call it. Reporting "part" is all you can possibly do in a journal article or brief symposium paper. It is a reasonable guideline for developing a full-blown study as well.

Do you remember Alfred's dilemma described at the end of Chapter 2? (I'll bet you do. We tend to remember material presented through anecdotes and personal asides.) Had Alfred come seeking my advice, I intended to ask if he had considered taking some manageable "unit of one" for a focus, some *portion* of the year's activities that would have allowed him to represent what he had fathomed from his extended data collecting without having to recapitulate the entire year. Might he have constructed his study around *one* student in the class rather than all of them; meticulously analyzed *one* day in class rather than every day; dissected *one* social studies unit from inception to post-test rather than trying to review all the units presented during the year; analyzed *one* critical event rather than regard everything that occurred as critical? Thus, to the advice of writing a study "bird by bird," there may be an acceptable alternative: Write only, or mostly, about *one* bird, in depth. Then discuss where that example fits in the broader spectrum. That has always seemed the more attractive alternative to me.

Returning to the zoom lens analogy, one way to keep a descriptive study manageable is to "zoom in" progressively closer and closer until your descriptive task is manageable, then "zoom out" again to capture the broader perspective. Like a viewer, a reader, too, needs context to know how the single case fits into some larger scheme of things. What can we learn from studying only *one* or *one aspect* of anything? My answer may seem facile, but I stake my career as a qualitative researcher on it: All we can!

REVIEWING FOR CONTENT AND STYLE

If I have given the impression that I take style to be more important than content, let me hasten to correct that idea. Content is paramount—what you have to say, not how you say it. Style is critical but secondary in writing anything; like data collection, it is necessary but not sufficient. An attractive feature about style is that it is amenable to discussions like this. No one can teach you how to write; lots of people can suggest ways you can write better. I have never met anyone who reads qualitative studies for style. What Clifford Geertz says of anthropological writing applies to qualitative research in general: "Good anthropological texts are plain texts, unpretending. They neither invite literary-critical close reading nor reward it" (1988:2).

For the most part, our inquiries are concerned with how other humans live their everyday lives. Everyday experience is common to us all; our studies should not be pretentious. There is a fundamental fascination with the way other humans live, and our accounts should bubble with the stuff of life itself. When they do not—when the accounts appear sterile and lifeless—how often is inattention to writing at fault? Our peculiar genius seems as often to take the life out of our studies as to celebrate the lives in them. Our opportunity is also our challenge: to portray real people doing and saying real things, as seen through the eyes of another human observer intent not only on helping us to see but helping us to understand.

A first step is to compare what you have in your draft with what you promised or implied in your original problem statement or Table of Contents. With a completed draft of the full account—including initial attempts at analysis or interpretation, no matter how rudimentary—the process of tightening begins in earnest. Tightening is the part of the writing process that I enjoy, although I do not suggest that working through successive drafts of a too-long manuscript is without agonies of its own. I try to express my thoughts clearly the first time. I do not write poorly simply to make revision more challenging. I take little delight in recognizing long, unnecessarily complex, and poorly formed arrangements of words in sentences or paragraphs that seemed adequate when I drafted them.

Worse yet, behind many such sentences and paragraphs lurks a poorly formed idea. I often wonder if my thinking is as convoluted as my writing. (If so, is the problem diminishing with experience or getting worse with age?) Thoughts seemingly satisfactory in the abstract seldom appear so crystal clear when rendered as text.

With an entire manuscript drafted, you may suddenly become aware that it is no small task to get description and interpretation to match. We carry in our heads rich overviews of our studies that we can never quite commit to print, a fact always more apparent in reading the work of others than our own. One sometimes wonders whether (other) researchers pay sufficient attention to what they have written. Does the narrative account support the analysis? The resolution lies not with faulting the analysis, but to ensure that the analysis helps to shape the descriptive account, just as the descriptive account provides the substance for analysis. I have suggested that writing description is a good place to begin, but that does not mean the descriptive account then becomes sacrosanct. Until a manuscript is in print, not a word you have written is sacrosanct. (You cannot be that cavalier with material quoted from your sources, but the choice and extent of such quoting are entirely yours.)

As the analysis takes shape, some material originally included may become superfluous, just as other sections may need to be fleshed out with more detail. Anthropologist Michael Agar offers this homespun description of the process:

> You learn something ("collect some data"), then you try to make sense out of it ("analysis"), then you go back to see if the interpretation makes sense in light of new experience ("collect more data"), then you refine your interpretation ("more analysis"), and so on. The process is dialectic, not linear. [Agar 1996:62]

The nexus between description and analysis in the writing is also dialectic, each facet informing its complement, each helping with the important work of reducing detail, maintaining the focus, and moving the account ahead. Where your descriptive elements are "thin" because your data are thin, try to exercise both candor and restraint. There is nothing wrong with sharing hunches or impressions, provided they are labeled as such. Nor can you be faulted for pointing out what additional data would have been necessary to support a generalization that you believe is warranted but are not yet prepared to make.

This critical "tightening" phase is also a time to look for needless repetition. Because our studies sometimes take years to research and months to write, we forget that they can be read in a matter of hours, even minutes. Sentences written weeks apart and revised days apart may be read moments apart. Astute collegial reviewers can be especially helpful in identifying repetitions that authors themselves are no longer able to recognize. How surprising to find a brief comment noted on your draft, "Didn't you just say all this about three pages back?" But how much better to receive gentle admonition from a colleague than to find a terse comment from an anonymous copy editor: "Redundant. Needs rewriting!"

REVISING AND EDITING

Writers writing self-consciously about writing sometimes distinguish between revising and editing, the former in reference to reviewing content, the latter in reference to style, correctness, and other detail. The distinction correctly sets priorities: Content comes first. But I wonder if the distinction also fosters a mistaken image of a two-step process. Editing presupposes something to edit, and that something is the essence of it, not simply a first step. Still, the implication that everything I write needs revision is a little harsh for my sensitive author-ego, even if it proves essentially correct. I find myself referring to manuscripts as revised *after* I revise them, but while actually engaging in the process, I prefer to think (and announce) that I am editing, whether I am performing a major overhaul or minor tune-up. I admit to revising only when I must change a format, make drastic cuts or revisions called for by an outside editor, or undertake a substantial rewrite, such as when I "de-dissertationized" my doctoral study (HFW 1964) for publication in 1967.

Most of what I call editing probably would be considered "revising" to anyone looking over my shoulder. You may recognize a subtle distinction you make between editing and revising when you assign your current working copy of a manuscript a new date, title, or code and relegate the earlier version to the archives. And, of course, you *should* occasionally save and set aside copies of your working drafts, so that you can track how the manuscript has evolved. That also allows you to go back to an earlier draft if you realize that some new improvement isn't an improvement after all or (my perennial fear) that you inadvertently press a key that obliterates your efforts.

Becoming Your Own Editor

"Editing" is a special skill and is not the same as writing. But it contributes to your knowledge of how words work and is a skill we can all practice in order to become better at it. Opportunities to edit present themselves in countless ways, and every researcher-cum-author should take advantage of them. One way to gain experience that seems particularly well-suited to academic and professional writing is through collegial review. I trust you recognize editing as help that can be *given to others* as well as help you seek for yourself. Providing the service of editorial review for colleagues not only gains you valuable experience but enlists you in our *collective* responsibility for the quality of our studies. Editing the work of others also affords opportunity to recognize desirable and undesirable practices in their writing that we are not always able to discern in our own.

If one's timing is right, graduate students polishing the final draft of a thesis are an especially receptive audience for editorial direction. The only caution is to try to avoid editing *for* them rather than helping them become better writers on their own. Because such writing is often (nay, usually) done under the pressure of complex power relationships and fast-approaching deadlines, well-intended efforts to be helpful can be perceived as intrusive instead. In that case, as with the analogy to assembling a wheelbarrow, perhaps all you can do is to make sure that budding authors have all the parts properly in place before they begin tightening. By all means, commend any sections that sparkle with insight and clarity.

Reading published reviews provides another means for keeping tabs on colleagues' writing and offers a 3-for-1 return on your investment of time. First, you get an overview and critique of new publications, a great help with the ever-pressing problem of "keeping up" in your field. Second, you get one academic author's reaction to the writing of another. Few reviewers can resist the temptation to comment on organization and style, although most of us like to think we are above such things. Third, you sample the reviewer's own style, seen in a disciplined piece of writing addressing the delicate business of collegial (and sometimes not-so-collegial) review. Book reviews are an underappreciated art form in academic writing. Read them. Write them. Pay no heed to anyone who insists that book reviews—whether published or electronic—"don't count" as scholarly contributions. Anything that makes you a more astute writer/editor contributes immeasurably to your ability.

Time—Your Silent Partner

When editing my own material, I experience a sense of diminishing return when I devote a too-sustained effort to the task. Manuscripts can always be improved (yes, this one too; how many more editions to get it "just right"?) and successive, productive editings are the way to get there. I also need distance from my words, however, lest I find that I am changing text without necessarily improving it. When time permits, I like to put manuscripts on a figurative "back burner" for a while, turning attention to other things before returning to editing. After a period of benign neglect, I can do a better job of strengthening interpretations; spotting discrepancies and repetitions; locating irregularities in sequence or logic; and discovering overworked words, phrases, and patterns.

Other Ways to Edit

When outside help is not available or time does not permit, I look for other ways to gain a fresh perspective. One is to edit from last page to first, attending to final sections rather than always beginning at the beginning.[2] Others include reading a manuscript aloud, reading a too-familiar manuscript in an unfamiliar setting, or reading a manuscript quickly, especially if all prior readings have been careful word-by-word ones. By simply changing font or format—something accomplished instantly with word processing—you can rearrange the spatial relationship of words on the page or screen and gain a different perspective on sentences previously fixed in your mind.

Sometimes I go through a manuscript in mechanical fashion to see if I can eliminate one unnecessary word from every sentence, one unnecessary sentence from every page. When editing directly on the screen, if the bottom line of a paragraph contains only one or two words, I rise to an implicit challenge to try to eliminate an equivalent number of characters somewhere within the paragraph to reduce the overall length of the manuscript by one line. I marvel at the experience of watching a paragraph *literally* tighten up on the computer screen!

Every sentence containing a form of the verb "to be" is a candidate for rewriting in active voice if I can see a way to do it. Often I cannot find a way to rewrite such sentences, which makes it all the more important to improve those that I am able to improve. I also launch "search and destroy" campaigns to ferret out overworked expressions and overused words as I become aware of them. My continuing rampage concerns the word "very," a *very* unnecessary word and habit. In their popular *Elements of Style,* Strunk and White note *rather, very, little,* and *pretty* as "qualifiers" and offer succinct advice: Avoid them (1972:65). Other unfortunate word habits that I have become conscious of are my overuse of "even" and "however." I've always written with too many "buts," but I have a hard time eliminating them. (Does that reflect a contrary nature?) I also seem to pick up, yet remain unaware of, some new hard-to-break habit with each writing assignment. For example, I have just (just?) discovered that I not only overuse the expressions "on the one hand" and "on the other hand," but that I frequently forget to mention that first hand altogether.

A bumper sticker imploring us to "Eschew Obfuscation" provided excellent advice for academic writers, but the slogan hardly rolls off the tongue. My current checklist of things to watch for in the final editing stages is short, but it usually takes more than one pass-through to catch them all.[3]

- unnecessary words
- passive voice, especially forms of the verb "to be"
- qualifiers, such as those noted
- overused phrases
- excessive use of anything, such as overuse of quotes, italics, parenthetical comments, and pet terms or phrases

Having discovered ages ago how (very) long it can take from the moment of first submission of a manuscript to finally seeing it in print, I continue to review drafts (even) after I submit them. (Remember: Only the final version counts!) I advise publishers of my edit-while-I-wait practice, noting that I will be ready with a clean and current draft the moment a manuscript is ready to go into production. Production can sometimes be delayed for a year or two (material submitted for edited collections seems to take the longest), and no draft can fail to be improved by periodic review during a waiting period. Should the period of delay become an extended one, references also may need updating. You may be both elated and dismayed at the number of additional references you discover *after* you send a manuscript off. Perhaps recent citations can be added at the last minute to make your study appear up-to-date, but remaining "current" is a game you cannot win. The earlier you suggest additions or changes to a manuscript already in production, the less resistance you are likely to encounter from an editor or publisher.

You may be able to strengthen your interpretation or analysis during the time for reflection prompted by unanticipated delays. Recognize that your most profound insights may not occur for years. As the epigraph to this book suggests, should you come to realize with the passage of time "how much the views on all points will have to be modified," you are nonetheless in good company!

On the other hand (OOPS!), don't drive yourself crazy tracking down new leads or trying to stay up-to-the-minute with what you publish. The new **electronic journals** are better suited for such reporting, and urgency seems a bit forced for descriptive accounts of the kind that most of us prepare. An invited chapter I once wrote for a prestigious audience of researchers suffered so many delays while the editor bullied and cajoled recalcitrant contributors that by the time the book was in print, my chapter intended as "state of the art" looked more like a historical review. The lesson for me was to edit with an eye for a substantial and enduring piece that could stand the test of time, rather than try to be "cutting edge" current. That stance may eventually prove to be our saving grace, for with the prospect of ever quicker means of publication, we recognize that our efforts do not have the same urgency as do hastily announced developments reported in certain other fields. With the prospect of relatively longer shelf life comes the responsibility to make our accounts both full and accurate.

FORMAL EDITORIAL ASSISTANCE

I have discussed the kind of help we can receive (and offer) in collegial review while a manuscript remains in our hands. Next I turn to what you can anticipate by way of formal editorial assistance, including "help" that is beyond your control.

One thing you can count on is that you cannot count on receiving editorial help with a manuscript submitted to a journal or publisher. An author of my acquaintance submitted the draft of a book-length qualitative study to a university press that seemed eager to publish it. She was aghast to have her draft returned *in copyedited form,* ready for final approval to go directly to the printer. What she had expected—counted on, really—was an editor's careful sentence-by-sentence examination to guide her own final editing. She had to make an agonizing decision (especially because this was her first book) to withdraw her manuscript, seek independent editorial help, and resubmit later in the hope that the publisher's enthusiasm would not wane. Fortunately, it did not. The alternative to seemingly rigid deadlines—in this case, at least—was to counter with a superbly revised manuscript.

Some time later, I had an opportunity to discuss my friend's dilemma with a representative from her press who appeared taken aback by the implication that the publisher offered no help. "We would have been happy to be more helpful by providing editorial service," she explained. "We weren't aware that it would have been welcome!" Authors are often reticent about soliciting help, concerned about giving any hint that they themselves feel their writing needs "help," especially should that result in a call for substantial revising or drastic cutting.

Matching editorial help offered with editorial help desired is an uneasy business under any circumstances. I am meticulous about editing. I followed that practice during the years I served as editor of a scholarly journal. I encouraged outside reviewers as well as our in-house staff (a part-time, semi-retired professional editor and several graduate assistants) to be equally rigorous. We penciled comments and suggestions on any manuscript under serious consideration, and we usually circulated a single copy so that each of us was privy to suggestions made by earlier reviews. This was in part for our mutual edification. We read each other's comments and argued among ourselves—in the margins of the manuscript, if we felt it might be of interest to the author—whether our suggested changes were always improvements. Other than with format requirements, however, we neither insisted on the changes we proposed nor guaranteed publication on the condition that authors make the changes suggested. Fresh draft, fresh review. When manuscripts came back revised—as they almost always did—we read them anew rather than compare them sentence by sentence with the earlier submission.

I was told that when we returned one manuscript, a senior colleague at another institution stormed out of her office and announced, "I haven't had anything marked up like this since I was a sophomore in high school!" What we received at the editor's office, however, was a gracious thank-you for our careful reading and a much improved draft that we were delighted to publish. On another occasion, a contributor reacted upon seeing his article in print, "I didn't know I wrote that well!" Truth was, he didn't. With our insistence and some specific editorial suggestions, coupled with his willingness to rework the material, the result was excellent. Because we insisted on better writing, we got it. I would be flattered to have an astute reader/critic someday discover and commend the caliber of writing in the journal under my editorship (*Anthropology and Education Quarterly*, 1983–1985), but that will never happen. Good writing does not call attention to itself, it only enhances what is written.

That academic writers make little use of freelance editors can be attributed, I believe, to frugality and a lack of precedent. There is no shortage of professional help available. (Check the Yellow Pages of your telephone directory.) We seem willing to invest great amounts of time at writing, and considerable sums on having the latest hardware and software available, yet nary a cent for editing. I do not recall ever seeing a line item budgeted for editorial assistance in a grant or project proposal, although the final product is often expected to be a publishable monograph or book rather than a technical report. An unstated but prevailing notion seems to hold that one's writing—like one's research—should be original, entirely one's own. Too bad, when writing so benefits from review by others!

Other arguments can be summoned against hiring professional editorial help, in addition to out-of-pocket costs that can run to hundreds of dollars. One problem is how to identify a "good" editor—the question of quality control in a field where virtually anyone who has ever written for publication or taken a few writing courses can feign expertise. Researchers also worry that the only help editors provide is with style, that is, that editorial consultants are uninformed on technical aspects and may not "understand" the material we place in their hands. Such an argument seems transparent when we claim that our objective is to help others understand by seeing through the perspective we provide. Editors ought to be able to help us accomplish that objective. Good editors do it without bruising tender author-egos, at the same time helping each of us to develop our own individual style. If only your best friends, your closest colleagues, or your students are your reviewer/critics, you may need to be reminded of the advertising slogan adopted for a popular mouthwash: "There are things your best friends won't tell you."

HOW DO YOU CONCLUDE A QUALITATIVE STUDY?

You don't. Give serious thought to dropping the idea that your final chapter must lead to a conclusion or that the account must build toward a dramatic climax. In the dichotomous thinking said to be typical of Americans, research is sometimes portrayed as either decision-oriented or conclusion-oriented. Clearly, some research is decision-oriented, but I am not sure that "conclusion-oriented" is a proper label for the rest of it. In reporting qualitative work, I avoid the term "conclusion." I also avoid the word "findings," for it seems to have a similar effect on reporting style by calling undue attention to details amenable to rigorous analysis rather than to the basic issues we may want readers to ponder. It all gets back to purposes. The more the problem seems to call for systematic data collection, reporting, and analysis, the more the research would seem to call for a quantitative approach. I do not work toward a grand flourish that might tempt me beyond the boundaries of the material I have been presenting, or might detract from the power (and exceed the limitations) of the observations themselves or what I tried to make of them.

Qualitative researchers seem particularly vulnerable to the tendency—and urge—to go beyond reporting *what is* and to use their studies as platforms for making pronouncements of *what ought to be*. A critical divide separates the realm of the observable from the realm of values, the good and the better. This is not a matter of simply taking a big leap. We cannot bridge the chasm between the descriptive and the prescriptive without imposing *someone's* judgment, whether originating from the people in the setting ("What we really need around here . . ."), from expert opinion ("If these people knew what was good for them . . ."), or from our own personal assessment ("On the basis of my extensive experience, I strongly recommend . . .").

There is an implicit evaluative dimension in all description. The antidote is restraint. The urge to lend personal opinion and judgment seems to become strongest when we begin searching for the capstone with which to conclude a study. You can recognize it creeping into your work (or, if you prefer, in mine) with the appearance of words like "should," "must," "need," or "ought."

There is nothing wrong with offering personal opinion or professional judgment. But it is vitally important to label it carefully and to search out and acknowledge its origins in your thinking. While you're at it, you might give some thought to why we feel duty-bound to come up with conclusions, and why the conclusions are supposed to be filled with cheery optimism. Anthropologist Ruth Benedict observed years ago that "American popular audiences crave solutions" (1946:192). As both producers and consumers of research, we need (need?) not only to recognize this collective penchant for closure but to recognize as well the corresponding urge it prompts in us to supply "satisfying endings."

How often today do we read about films produced with tentative and multiple endings while producers argue over, or focus groups help decide, which finale is likely to draw the biggest box office? Every article in our weekly news magazines, every report of on-the-spot TV coverage, has its dramatic tagline. The endings for qualitative studies do not have to be dramatic; they need only to be well-suited to the occasion. In a dissertation, nothing more may be necessary than a sentence or two tucked into the last paragraph of the chapter dealing with interpretation. Academic restraint predominates in journal publication and the more scientifically oriented monographs. Books seem to accommodate, and to demand, more author input, even a bit of flair in a closing statement. Thus, one's audience remains a key factor, as does one's stage in a professional career.

My recommendation for anyone new to academic writing is to work toward a conservative closing statement that reviews succinctly what has been attempted, what has been learned, and what new questions have been raised. Do not abandon a detailed case study in a last-ditch effort to achieve a grand finale. *It is not necessary to push a canoe into the sunset at the end of every presentation.* Recognize and resist the temptation to offer dramatic but irrelevant endings, or conclusions that raise issues not addressed thoroughly in the research. Beginnings and endings are important; they deserve extra attention from the author because they tend to receive extra attention from the reader. Look for ways to make them better without letting them become more dramatic.

Rather than striving for closure, see if you can leave both yourself and your readers pondering the essential issues that perplex you. In time, you may understand more. As noted, only in a 1989 afterword to *A Kwakiutl Village and School* did I find an adequate way to bring closure to the study I first presented in a doctoral dissertation completed in 1964. Lapsed time: a quarter century!

Nevertheless, also be warned that where and how you might prefer to bring the account to a close may not go far enough to satisfy critics. Commercial publishers and experienced editors like to remind their potential authors that they "know their audiences" and may insist that you offer more by way of a summation "because that's what our audiences expect," even while they insist that you shorten your manuscript. If you argue (and no doubt you would like to, whether or not you actually do) that the case stands by itself, or that the meaning of your research is not all that clear, then you may be pressed (by a wily editor, a granting agency, even a dissertation committee) to state what *you* learned, or to reflect on what *you* think it all means.

ALTERNATIVES TO "CONCLUDING" A STUDY

Some alternatives to writing a formal conclusion include summaries, recommendations and/or implications, or a statement of personal reflections. Any one or a combination of these may satisfy the need for closure without tempting you to go too far, losing your audience just as the final curtain descends. Each of these alternatives raises questions about purposes and opportunities in qualitative research and about your intended audience.

Summaries. An objective summary can provide a careful, restrained way to end on a strong note. It allows a review of what you have accomplished in terms of your original statement of purpose. It also provides opportunity to anticipate critical reaction—the well-intended kind, not your Worst Nightmare Critic—by pointing out shortcomings and discussing how, now that you are older and wiser, you might better have conducted the study. Restrict your comments to summarizing what has gone before. A summary is not the place to startle readers with important information that might, and should, have been introduced earlier.

Do not be tempted to introduce an interpretive emphasis that gives a totally new twist. A summary provides opportunity for repetition and emphasis to ensure that your message gets across, if indeed there was a message. But if your summary is also an editorial, join the title with some other word that signals your intent, such as Summary and Discussion, or Summary and Reflections (see below).

If the idea of summarizing appeals to you, consider going a step further by providing brief summaries *throughout* your study, as I was asked to do for this revision, rather than saving everything for a Grand Finale. Academic authors could make better use of chapter or section endings if they reserved them strictly for summarizing. Too often, sections labeled as summaries are devoted to anticipating what is coming next, rather than fulfilling the promise of a succinct review of material previously covered. Introductions, as the word suggests, belong at the beginning of each new section, not at the close of the preceding one. Concise and well-written chapter summaries can provide a sort of running "box score" for stating how things stood at the beginning of the chapter and reviewing important new information and insight. Summaries should help everyone remain on target, author and reader alike.

Recommendations/Implications. A frequent practice for resolving the how-to-conclude question is to prepare a final section or chapter that couples a brief summary with recommendations or implications. Whether boldly to offer recommendations or more tentatively to tease out implications depends on the nature and purpose of the study, its intended audiences (e.g., policymakers insisting on recommendations), and the posture (and status) of the researcher (e.g., dispassionate observer, consultant, critic, advocate).

A call for recommendations can put the more objectively oriented researcher in a bind. One would like to present the case so thoroughly and in such a well-contextualized way that the reader has the same basis for making judgments as the researcher—and thus the researcher is relieved of responsibility. Still, if you have devoted extensive attention to a problem or setting (e.g., Why don't the children of this ___ [ethnic minority of your choice] group perform better in school? What steps could be taken to curb the ___ [social problem of your choice] among these teenagers? What might be done to improve the _____ [socially desirable goal of your choice] among members of this group?), it is not unreasonable for sponsors or concerned readers to expect some helpful reflections or advice. And it certainly doesn't hurt to point to whatever is being done well, to counter the often negative tone that our studies acquire as we describe the consequences, intended or unintended, of programs designed to "help."

Descriptive studies can be maddeningly ambiguous. For the busy practitioner or policymaker, the bottom line is always "So what?" or "What's to be done about it?" For such audiences, a researcher's efforts to convey nonjudgmental objectivity is more likely to be perceived as an academic cop-out than a laudable research stance. We may prefer not to be pressed for personal reactions and private opinions, but we must be prepared to offer them. One way to share this responsibility is to outline the additional information or insight a researcher would need in order to pose a final solution, offer recommendations, or render the judgment requested.

Treated too cavalierly, or brushed aside with an unbecoming modesty ("Oh, we couldn't possibly say anything about that—we don't know enough yet"), the too-humble-to-be-helpful approach can be a cop-out. Yet the very act of pointing to elements the researcher feels he or she has not understood, or that seem poorly defined, may help uncover inherent ambiguities. To the question, "Why don't you tell us how to make this program more effective?" a researcher might have to reply, with a discomforting but not altogether unlikely explanation, "Because I have been unable to get a clear sense of what you are trying to accomplish."[4]

Another way to offer help—although it, too, can lead to discomfort and denial—is to identify **inherent** tensions and paradoxes. The manner in which people go about things often produces a different, and sometimes opposite, effect from what they intend. Anthropologist Ray McDermott provided an instance of this in his microcultural analysis of differences in the organization of reading instruction for the "top" and "bottom" pupils in a first-grade classroom. He observed how the top-performing readers practiced their reading skills while the bottom readers rehearsed classroom protocols appropriate to their niche as "poor" students (McDermott 1976). Although such an observation and interpretation offered by an outsider might bring little joy to a dedicated but harried first-grade teacher, comparable paradoxes permeate formal education whenever the quick students get the lecture and the slow only get lectured. Human social life is filled with paradoxes in which the consequences of behavior produce an effect opposite to what we wish to achieve, something that a detached outside observer is more likely to detect than is an engaged and committed insider.

A third way to offer help is to identify alternatives to current practice—or alternative solutions to current problems—and then systematically to examine the possible consequences of each alternative. In this way, the analytical skills of the researcher can serve not only as a potential resource but also as a model for others who may someday conduct inquiries of their own.

The ideal extension of this approach is that change agents such as nurses, law enforcement personnel, social workers, teachers, and so forth not only should be collaborators in research but ultimately should *become researchers* who conduct studies among their own clients (see, for example, Mills 2007). Our problems with data overload should help us appreciate why such an idea, one that sounds so "right," is often impractical. To an even greater extent than do researchers, practitioners must "get rid of"—which in this case means ignore—massive amounts of information before they can get on with their appointed tasks. To know more may hopelessly complicate assignments that are already hopelessly complex.

Drawing *implications* is similar to making recommendations but allows the researcher to remain more distant and contemplative. Identifying *possible* implications may offer an oblique approach in which questions are raised rather than solutions proposed. When one is addressing specialist audiences (e.g., practitioners, administrators, policymakers)—audiences whose members may not take kindly to boldly stated advice based on a neophyte researcher's modest study of perhaps only a single case—some way to convey that tentativeness seems warranted.

When our intended audience consists solely of research colleagues, I think it sufficient to conclude with a statement summarizing what has been learned and what appear to be the next steps in an ongoing process of inquiry. But we like to think we do more than simply talk to ourselves. If asked, we must be prepared and willing to say more, to offer what help we can.

We also can do a better job of inquiring into the kind of help that practitioners want, or make clearer the kind of "answers" we are in a position to render. One of the intriguing questions constantly before us—our own professional paradox—is why social research has so little impact. How often do we find ourselves scratching where it isn't itching? We do not give sufficient attention to the impact of our research efforts and the related question of whether that is exactly the influence we want to have. We agonize over that issue as a global one; perhaps we would be more convincing if we addressed it case by case.

Personal Reflections. I welcome the prevailing mood that encourages researchers to be candid and "self-reflexive" about the fieldwork experience. Unquestionably, the fieldworker is likely to be the individual most affected by the experience. If you close on a note of personal reflection, keep the subject(s) of your study the focus of your reflections. The more you feel an urge to step into the spotlight, the more carefully you should distinguish your personal reflections from the observations on which they are based, especially if your presence and feelings have otherwise been muted. If you have maintained a presence all along, you probably have had (or made) opportunities to share your personal reflections, and you probably have said enough.[5]

SUMMING UP

- In and of itself, the relatively greater length of qualitative accounts should not be of primary concern. Providing an adequate descriptive basis calls for detail.

- Unnecessary length, or the inclusion of seemingly tangential material, on the other hand, is distracting, leaving the reader to wonder if an author has lost the way or is telling stories for his or her own sake rather than to achieve a purpose.

- Attending to sentence structure offers a first step to the kind of tightening that so improves writing. Phrases like "in and of itself," or "on the other hand," in the points immediately above, can be edited out in the interest of an economy of style.

- Unless there is some compelling reason for presenting long interview protocols in an informant's own words, or drawing long quotations from the work of others, paraphrase and/or edit to lend emphasis to the material that you do quote.

- If cutting words per page or pages per chapter isn't sufficient to reach a desired (or imposed) page limit, consider deleting entire sections, even entire chapters, leaving some topics to be taken up elsewhere.

NOTES

1. For more discussion on the role and importance of visual display, see Tufte (1983, 1990). For more on creating charts, see Wallgren et al. (1996).

2. You may have discovered that making corrections from a hard copy to the screen goes faster if you work from the back to the front of an article or chapter. Text remains exactly where it appears on both hard copy and screen, making it easier to locate any changes you have noted.

3. My checklist prompted a recollection of the advice editor C. Deborah Laughton received from her writing teacher, Isabelle Ziegler, years earlier:

> Nouns are good,
>
> Verbs are better.
>
> Adjectives sometimes,
>
> Adverbs never!

4. See HFW (1983b) for an example of efforts to describe an ambiguously defined community development project.

5. See the discussion of two contrasting styles of researcher-reflective reporting, the Confessional and the Impressionist, in Van Maanen (1988), Chapters 4 and 5.

CHAPTER SIX

Finishing Up

I love being a writer. What I can't stand is the paperwork.

—*Peter DeVries*

You only thought your work was finished! If it is going to be published, there are a number of choices still to be made. Here are most of them.

It is not only a courtesy and good politics to be familiar with the format of a journal or monograph series for which you are preparing a manuscript, it is absolutely essential that you submit material in the manner requested. You may be surprised to discover a wide range of practices and preferences from one publisher, one professional field, and one journal to the next. Practices often vary dramatically from one editor's tenure to the next with the same journal, even when policy statements remain the same. It is the business of an editorial staff to establish the kind and extent of uniformity desired and to ensure that your manuscript fits within certain parameters. To accomplish this, you may face some unexpected writing and some unanticipated decisions. You will also find that many important decisions have already been made for you. These are among the kinds of finishing touches reviewed.

Major Parts of a Book (and Most Articles)

The materials that precede and follow the main body of text in a book are known by rather unimaginative labels: **front matter** (also "preliminaries") and **back matter** (or "end matter"). Most of the same "matters" must also be attended to in journal articles, reports, and chapters in edited collections, albeit in abbreviated fashion. These topics are discussed in the order in which you may think about or be requested to attend to them, rather than where you decide to place them. Some of these statements are intended to draw attention to you and what you have to report; some are intended to make what you have written easy to search, and some simply allow publishers to maintain a reputation for quality and consistency. As to both their inclusion and their location—should you have any choice in the matter—you may want to consider some possible alternatives that I will propose.

Finishing up the **front matter** requires decisions about your title, dedications, preface, introduction, foreword, acknowledgments, and table of contents. It may entail a section "About the Author"; preparation of an abstract; and, for journal articles, identifying key words, descriptors, or index words. There are also final decisions internal to the manuscript that must be made if, to this point, they have been in flux: whether and where to use footnotes, endnotes, or neither; whether and where to use tables, charts, and diagrams; and whether or how to use photographs and artwork.

Finishing up the **end matter** includes attention to references, and, for book-length works, decisions about whether to include appendixes or supplements, and whether the manuscript warrants, or the editor insists on, an index or a glossary. That takes care of the formal end matter, but that does not end matters for you as author. Still remaining are your responsibilities for responding to queries from your copy editor and for page proofing. If you are preparing your first study for publication, you may long for the day when any of these problems—even the final decision about a title—are the problems that concern you. That may also lead you mistakenly to believe that they can be left for later. Giving some thought to them as the work progresses can help you avoid making hasty decisions or having to do last-minute chores—such as checking references and getting them into the proper format—that are more efficiently handled as the manuscript is developed.

THE TITLE

A shorthand title may prove adequate in the early stages of a project. The working title of every paper, article, or book you write, and the date of your current draft, should appear on every page. If your working title encapsulates your problem statement and helps to keep you ever mindful of focus, so much the better. Be thinking about possible final titles from the beginning and jot down ideas as they occur. During the long interim between the start-up of a project and a completed first draft, the title is one of the few tangible aspects you can share that both announces and summarizes your study. In a reflective article "From Title to Title," Alan Peshkin, whose several book-length studies on aspects of American communities and education provide excellent models of qualitative research, described how, during the course of a field study, the evolving sequence of possible titles reflected his thought process as he continuously refined his research focus (Peshkin 1985).

Selecting a title is serious work, but it can also be fun. A common practice in scholarly writing is to assign what amounts to a double title. As a consequence, two long, independent, often seemingly unrelated subtitles, joined by a colon, may be attached to even the shortest of articles. One of these titles may be creative, even catchy. The "catchier" it is, the greater the need for a subtitle that gives a clear indication as to content.

Some of my early favorites among such titles are Suzanne Campbell-Jones, *In Habit,* with the informative subtitle *A Study of Working Nuns* (1978), and Shari Cavan's *Liquor License: An Ethnography of Bar Behavior* (1966). Janet Spector's *What This Awl Means: Feminist Archaeology at a Wahpeton Dakota Village* (1993) presents a title that promises not only some serious digging but a lively read.

I caution against being too cute. Titles can come back to haunt you and may detract from your purpose. If the lighthearted part of your title is on the clever side, its complement, the subtitle, should accurately describe the nature of your work. From firsthand experience, I also advise against using unfamiliar words in a title, especially place names about which the pronunciation is uncertain. My first book title included the word *Kwakiutl* (HFW 1967); a later book took the name *Bulawayo,* a city in Zimbabwe, for part of its title (HFW 1974a). Those names conveyed important information, but I discovered that many readers avoided the them—and thus never referred to either book by title—preferring not to stumble over an incorrect pronunciation. An article or book with a title that one cannot pronounce is not a likely candidate for becoming a topic of conversation.

My partner Norman and I are credited by Ron Rohner for suggesting *They Love Me, They Love Me Not* (Rohner 1975) as the title for his then-newly-completed manuscript, but our creative inspiration would have done a grave disservice without the complementary subtitle: *A Worldwide Study of the Effects of Parental Acceptance and Rejection.* Similarly, *Teachers Versus Technocrats* (HFW 1977) proved an effective title for a case study of the dynamics of educational change, but it sorely needed its subtitle, *An Educational Innovation in Anthropological Perspective,* to bring it to the attention of its intended audiences. Granted, either of these subtitles is a mouthful, but they helped inform potential readers and signaled fair warning as to their serious orientation. I like to chide academic colleagues about their long titles, but we are not alone. The complete title of a Charles Dickens classic, usually referred to only by the name of its central character, is *The Personal History, Adventures, Experiences & Observation of David Copperfield The Younger of Blunderstone Rookery (Which He never meant to be Published on any Account).*

Should a Hollywood studio approach me with the unlikely possibility of making a film based on this monograph, I'll cast about for a much snappier title (something like *Romancing the Keys?*) to replace the cumbersome *Writing Up Qualitative Research.* Until they do, however, my conscience is clear. The present title succinctly and accurately conveys enough about the contents to hold its own in the marketplace of ideas. It is short, but not too short to communicate. Exceedingly short titles may render a disservice. One that comes to mind is Gregory Bateson's succinctly titled *Naven* (Bateson 1936). Although eventually recognized as an "eccentric classic" (Geertz 1988:17), the book's title only compounded the obscurity in which it remained shrouded for more than two decades. But it would be hard to top *Rc Hnychnyu* (Salinas 1978) for a title guaranteed to scare off any but the most dedicated student able to recognize that the account deals with the Otomí people and language.

Computerized databases have added another reason for including critical locator words in a title or subtitle, especially for book-length works. If important identifying words do not appear in the title, the work will not "come up" during a computer search and may not attract attention in a publisher's catalog. The "cute" alternative title suggested above, *Romancing the Keys,* is a good example of a bad example. And think how *Bird by Bird* might have been cataloged on an electronic bibliography had Anne Lamott not added the subtitle *Some Instructions on Writing and Life!*

FRONT MATTER

It may seem a bit obvious that front matter goes at the front of a book. Except for a Table of Contents, augmented perhaps with an executive summary, I am not convinced that loading up with customary "front matter" baggage is a great idea. Let me review some of these "matters" with an eye to placing them elsewhere (i.e., at the back of the book instead of the front) or eliminating them altogether. This may be another of those times when you need to put yourself in your readers' shoes. Readers are anxious to get to the content of your study; this is no time to get in their way!

Dedication. Academic authors sometimes go overboard with the well-intended but subject-to-abuse practice of dedicating works, particularly works of limited scope or modest appeal. My suggestion is to *acknowledge* the help and support of others (including your spouse and offspring who, it would seem, somehow *were* able to convey the isolation they suffered during the prolonged period you devoted to writing) rather than express gratitude or affection in a dedication. I think that dedications should be reserved for the finest of works and the most special of people. With lots of special people in mind, I have been able to resist encumbering anyone with a dedication thus far. It's always tempting, but I intend to hold out a bit longer. Your call, of course; if you insist on indulging yourself, keep the dedication simple.

Preface. Prefaces, like any prefatory statement, serve the important function of setting forth the purposes and scope of what lies ahead. They give the author a personal opportunity to invite the reader to come in for a closer look, with the blessing of the publisher, who probably views this as an opportunity to promote the book to a potential buyer. If you originally submitted a formal prospectus with the hope of gaining the publisher's interest in publishing, you might think of a preface as a sort of "second wave" prospectus written to attract and appeal to a broader audience, now that the work is in print.

Yet something seems to happen to many authors when the time (finally!) comes to write a preface. Written last, and often in a style too revealing and personal for an author we have not yet met, prefaces are placed where they will be read first. As author, you may wish you could address your reader in a direct, personal way about your work, but if that is your purpose, I suggest you do it later. Save your reflections or confessions, even your acknowledgments, until readers know more about your study and *may* appreciate an opportunity to know more about you. In lieu of a preface, consider *concluding* with an epilogue or a literal afterword, or add either personal reflections or "A Final Note" as a postscript.

If you do write a **preface**, keep in mind that it is the book or monograph that you are introducing, not yourself. If you already accomplish this in an introduction (see next), consider whether you really want to add what may result in little more than another, albeit shorter, one. Don't be tempted at the last moment to upstage the whole writing project that has been consuming you. Browse the works of authors you admire to see how, or whether, they began their accounts with a preface and whether, in your judgment, the preface really contributed anything.

Introduction. Preparing an *introduction* separate from the text presents another temptation for academic throat-clearing. I recommend against writing a stand-apart introduction, for it is likely to be little more than a longer and more formal preface in disguise. Chapter 1, page one, is where the reader should meet the author, and nothing should stand in the way of their meeting immediately. Of course, your first chapter can be *titled* Introduction, or an introduction can serve as Chapter 1—in either case, the author gets right at the substantive matter of the text. What is written later by way of reflection can appear later, rather than be allowed to distract or detract. If more explaining is necessary, the introduction itself probably needs to be rewritten.

Foreword. Perhaps you (or your publisher) would like someone else to do that "throat clearing" for you. There are obvious advantages in bestowing that honor, and it is not unlikely that whoever is invited to write a foreword will reciprocate by lauding your work or otherwise attesting to the importance of your study. But you cannot anticipate for certain whether someone will rise to the occasion and do both you and your study good service. Soliciting a foreword carries a bit of a risk.

In the interest of getting the reader to the text as quickly as possible, my general recommendation is to dispense with both an introduction and a foreword. But there are exceptions. In a monograph series, it is likely that the series editor(s) will want to introduce each volume, so the foreword (or Editor's Introduction, as it was labeled in the first edition here) is assigned rather than solicited. And none of us is adverse to having someone say a few kind words about us or our work or offer explanations better written on our behalf than by ourselves.

In one earlier study (HFW 1974a), I included both a foreword *and* an introduction, each authored by a different individual. The study dealt with urban African drinking, and I felt that it would be good for the book, and for me, to be introduced by a recognized scholar on drinking behavior. I also felt a huge debt to Hugh Ashton, the anthropologist-cum-administrator who had made my study possible and whose blessing I wanted to secure for the completed work. I was pleased that *both* individuals invited were willing to prepare statements by way of introduction. Subsequently, there have been other times when I wanted someone else's words to validate my own. That is what these devices can do. Think of the decision about whether to include them as a strategic one in which the intrusion should be genuinely needed and warranted.

Table of Contents. I have extolled the virtues of preparing an early draft of a Table of Contents as a valuable tool, not only for subsequently organizing data but for organizing the field research as well. In the final stages of preparing a book-length study, the Table of Contents, with its chapter headings and subheadings, needs careful review in terms of appropriate titles, parallel treatment of like categories, and the sequence for unfolding the account. I recommend that, to make a critical appraisal of the contents, you pull out and list separately all headings and subheadings and examine them in relation to each other. Do they provide an adequate structure to hold the account together and a workable sequence for developing it? Authors of briefer statements (articles, chapters in an edited volume, research notes or brief communications in journals) should recognize that their headings and subheadings constitute an implicit Table of Contents deserving the same critical attention.

Preparing a formal Table of Contents is another occasion requiring decisions about level of detail. There is no universal formula, but trade-offs between as-much-as-possible and as-little-as-possible are fairly obvious. If you do not provide something by way of an abstract or a more explicit guide for your readers, the Table of Contents offers the only overview of what you are presenting, except in rare cases when a back cover or book jacket provides summary information.

As a general guideline, the more concise the Table of Contents, the better. When contained on a single page, the Contents serve as both an outline and a reader's guide. The problem with a brief, eye-catching, bare-bones Table of Contents is that chapter headings may not convey an adequate sense of the scope or depth of your study, particularly if you employ conventional chapter titles from the standard I-H-M-R-D sequence (Introduction, Hypotheses, Method, Results, Discussion). I hope you feel free to make your chapter titles more informative and more interesting than that. If you do not include an index (dissertations never do, and neither do many qualitative/descriptive studies), the Table of Contents provides the only guide for anyone trying to locate material within the study. That makes the case for providing a detailed one.

How to decide between a succinct Table of Contents and an elaborated one? You probably won't be given a choice; the publisher's preferences ordinarily prevail. But if you do have a choice, this is an instance where your purposes and your intended audience can inform your decision. "More" may be better, sacrificing elegance for thoroughness to convey the depth of your inquiry. In dissertations and unpublished reports, a detailed Table of Contents is not only appropriate but essential. As a compromise between too much and too little, consider making your chapter titles as descriptive as possible and then expanding individual chapter descriptions to an overall limit of what will fit on a single *printed* page.

Among several books near at hand as I write, it appears that monographs and shorter, single-authored books tend to observe the one-page format. Textbooks and edited collections have Tables of Contents that may continue for several pages, staking an implicit claim to comprehensiveness. Should writing and editing prove your métier and you someday find yourself author or editor of a huge compendium, you might do what Russ Bernard did for his comprehensive edited text *Social Research Methods: Qualitative and*

Quantitative Approaches (Bernard 2000). He provided readers with a "Brief Contents," consisting of two pages, followed by an expanded (11-page) "Detailed Contents" giving the breakdown of each subheading within the chapters.

Front Matter: Further Possibilities

Acknowledgments. In *The African Beer Gardens of Bulawayo,* I included the acknowledgments with the other front matter. It hadn't yet occurred to me that I didn't *have to* put them where everyone else did. Now that the idea has occurred to me, I usually place my acknowledgments (also *acknowledgements,* the preferred British spelling) at the back of a book.

I received an early lesson about the importance of acknowledging others from George Spindler. The Spindlers were among my first house guests after I completed doctoral studies and had accepted a full-time academic appointment at the University of Oregon. The evening they arrived, I eagerly shared with them a draft of a paper I had been invited to write, tentatively titled "Concomitant Learning." Spindler arose early the next morning, but to my disappointment, I found him looking through materials *he* had written (my library contained most of them) rather than reading my new draft. He had already read and "enjoyed" my article, he assured me, but he expressed disappointment at my failure to credit him as source or inspiration for the *concept* of concomitant learning that provided my title and rationale. He had been searching through his own published pieces for the citation that I might have made. "But you've never written about it," I explained, reaffirming what I already knew and he was beginning to realize. "I got the *idea* from you, but you only suggested it in seminar discussions. There was no publication to cite."

Technically (and luckily) I was correct, as his search had revealed. That wasn't the entire lesson, however. "No matter where or how you encounter them," he counseled, "always give credit for the sources of your ideas. It's so easy to do; so appropriate to good scholarship, . . . and so appreciated." Never again have I limited my acknowledgments only to people whose ideas are in print.[1] And I, too, have "so appreciated" that courtesy when extended to me!

As with much of the front matter, however, acknowledgments can be placed at the end rather than at the beginning of the text. They are less distracting there, and by the end of the manuscript, readers should better understand what is being acknowledged. A preferred form for general acknowledgments (e.g., inspiration, reviewers of early drafts, even particularly helpful anonymous reviewers) in many scholarly journals is to contain them in a first (and unnumbered) endnote, followed by numbered endnotes that, among other things, may include additional acknowledgments or permissions.

It is the traditional *place* assigned to acknowledgments that I object to, not to the *practice* of acknowledging. I now make an effort to share as much credit as I can without compromising confidentiality necessitated by the reporting itself. An idea borrowed from novelist James Michener is to keep lists of those who help at each *stage* of the work and to acknowledge their contribution in the same sequence. I keep a log of the names of those who assist in any important way during the course of a study or preparation of a manuscript, not just those involved with the final draft. It took seven paragraphs to acknowledge the help I received with the research and writing for *Teachers Versus Technocrats*. I do not recall anyone faulting *that* section as overwritten.

About the Author/About the Book. To whom do you think falls the task of preparing those brief but glowing sketches that accompany articles and chapters or appear on back covers and dust jackets? Chances are it will fall to the author. Should you be asked to prepare a "bio," accept the assignment as another opportunity to recruit readers and to establish your authority to do the kind of research addressed in your report. What you say of yourself should link the study to your experience, your expertise, and your career; this is not the place to share the enjoyment you derive from gardening or listening to classical music. Of your experience and past accomplishments, be specific and to the point. I appreciate authors who cite their *relevant* previous works by year and title, rather than those who claim to have published "several books and numerous articles on a wide variety of topics."

Sometimes, a parallel statement "About the Book" accompanies the author's sketch described above. Such a statement is not intended as a reader's guide as much as a short history of how the book came to be written and its place in the author's career. If offered in lieu of an abstract, it probably needs to be up front as a preface, although my preference is to join these short pieces, *About the Author* and *About the Book,* and to include them with the other back matter. The reader who seeks such information can locate it easily enough. My rationale for placing all such material at the back is that readers are likely to be interested in knowing more about the author, or the role of the study in the author's career, or the people who helped along the way, *after* reading and assessing a book's contents than before. Attention should remain on what the author has to say rather than to his or her credentials. If you feel, or for marketing purposes your publisher insists, that this information serves a vital function by way of introduction, you should comply (as I have here). In dissertations, this function is performed by presenting a candidate's abbreviated vita. If an author wants to say more, it can appear in the text itself.

The Abstract or Précis. It should be apparent that with the exception of a Table of Contents, I am not wildly enthusiastic about packing the front matter with the material usually placed there: dedications, prefaces, introductions, acknowledgments. But one possible category appears too infrequently: some sort of executive summary to call the attention of potential readers to what you have to report, even perhaps to suggest how the "busy" reader (i.e., one whom you suspect will read only hurriedly, if at all) might approach the reading.

The *abstract* is the most common form for presenting such information. Professional journals require that an abstract—not to exceed a clearly specified number of words—be included with every major article submitted for publication. Publishers some-times make a similar request of an author submitting a longer man-uscript, although they usually end up writing such copy themselves because abstracts are a crucial marketing tool. That may explain why glowing previews characteristic of book jacket blurbs show a marked contrast with impersonal author-abstracts found in profes-sional journals. Most academics can't get pumped up enough to "pitch" their own books, although most of us are not adverse to having someone do it for us.

Guide to the Reader. An abstract, précis, or executive summary can prove helpful to readers and thus is indirectly of benefit to the author as well. We should encourage wider use of anything that improves readability. An author can be of even greater service by supplying a Guide to the Reader, not a bad idea for a scholarly monograph that might otherwise be overlooked because of its very thoroughness. If it is truly a *guide*—rather than an abstract in disguise—it also may indicate where to locate specific subtopics *within* the text.

The earliest use of an executive summary in qualitative research that I can recall appeared in a monograph by Murray and Rosalie Wax and Robert Dumont, Jr., *Formal Education in an American Indian Community,* originally published in 1964 and reissued in 1989. The entire study, only 126 pages in length (including 11 pages of appendix), stands as a model of field research and succinct reporting, particularly for its effort to reach an audience of practitioners. Yet even with so brief a monograph, the authors immediately catch a reader's attention with a three-paragraph (double-spaced, even on the printed page!) "Guide to the Reader" to convey the gist of their message to those they fear might not pause long enough to discern it for themselves. The first paragraph of their Guide appears here as Figure 6.1.

Note the special reading assignment for "skeptics and critics" in the final sentence. That caution marked a recognition by the senior authors (sociologist Murray Wax, anthropologist Rosalie Wax) that their qualitatively oriented approach—not well recognized outside their respective disciplines at the time—would be subject to scrutiny by methodologists, although probably of little concern to busy practitioners. Alas! This summary seems to have been one of too few occasions when someone made use of an executive summary in a descriptive study. It is an idea whose time apparently still has not come, although it remains a familiar feature of report writing.

Key Words, Descriptors. When submitting an article to a professional journal, you will be requested to supply some key words or descriptors to accompany it. As a journal editor, I was surprised at

Figure 6.1   Example of an Effective Executive Summary

A GUIDE TO THE READER

Those who must skim the pages of reports as they run from crisis to meeting to office are advised to turn to the chapter titled "Summary and Recommendations," which has been written with them in mind. Readers who wish to examine a picture of a contemporary Indian reservation and who are indifferent to the preliminaries of a research investigation are advised to turn to the second chapter, titled "Ecology, Economy, and Educational Achievement." Skeptics and critics will want to read not only the first chapter ["Perspective and Objectives of this Research"] but also the Appendix ["Research Procedure"] before proceeding into the heart of the text.

Source: Wax, Wax, and Dumont (1964:v). (Titles in brackets added.)

how little thought authors seemed to give to a standard request for index words. It was the practice of our journal to include several key words at the beginning of each article and subsequently to use those words to compile an index for the volume year. Too few authors seemed able to put themselves in the position of the reader searching an index to locate relevant materials. For the *Anthropology and Education Quarterly* as an example, "anthropology and education" was not particularly helpful as an index topic. Choose words and phrases that communicate your research *problem* or research *setting,* rather than your fieldwork techniques. Think how little information is conveyed using "participant observation" as a locator term.

The Poster Session as a Form of Abstract. The executive summary has a counterpart in the idea of the poster presentation at professional meetings. "Poster sessions" have become increasingly popular as an alternative form for presenting information, especially at conferences that must accommodate large numbers of presenters. The popularity of such sessions may be greater among program organizers than among presenters, although certainly there are researchers who dread the thought of having to prepare and present a formal paper. What the poster session entails is for the researcher to prepare a visual display—including printed text, maps, diagrams, photos, artifacts—that summarizes the problem addressed, how it was researched, and the outcomes of the investigation. Typically, the researcher is on hand in person during a specified period when the poster is on display, ready to interact with interested viewers who circulate among a number of such exhibits.

The invitation to participate in a poster session ought to be regarded as a welcome exercise for qualitative researchers. Anyone who has participated in a poster session might contemplate how something similar would be a useful supplement to a written report. Should your wish to present a paper at some future conference be met instead with an alternative assignment to a poster session— something especially likely to happen to neophyte researchers—take the opportunity to explore how succinctly you can convey the essence of what you have been up to, what you have learned. You can't just post a copy of the paper you intended to present; if you need a long explanation, prepare a supplementary handout that gives more detail. Use the space allotted to provide a visual overview of what you have done, in the same way that newspaper headlines are meant to entice us into wanting to learn more. Remember to feature your *name* and your *contact address,* as well as highlighting your statement of purpose (unless you are planning to make a career of remaining obscure). Of course, your name is not yet likely to be linked with the research you are reporting—that's why you're there, remember!

Dissertation Abstracts. Graduate students face an unexpected (and, catching them by surprise as it often does, not a particularly welcome) task when informed, *weeks* before they are scheduled to defend and subsequently to submit the final faculty-approved disk or copy of their dissertation, that they must submit an abstract for *Dissertation Abstracts* in final form. Today, these are filed electronically and are required in virtually all institutions of higher learning. For years, the collected abstracts have been published as bound volumes circulated nationally and internationally. This is the *only* widely circulated announcement that will ever appear about most dissertations. Too often, writing this brief but important statement is left to the last minute, at which time the author dashes off a hasty synopsis that needs instead to be concise, highly informative, and carefully written.

Having to encapsulate one's major professional preoccupation of the past months—or years—into the inviolable word limit of an abstract for a journal, or of one's dissertation study for *Dissertation Abstracts,* can seem like the last straw. Fortunately, it *is* about the last straw, a signal to celebrate that a major effort is finally nearing completion. As with anything you write, give time and thought to preparing your abstract, review it editorially, try it out on others, and ask someone to read it aloud to you. An abstract affords a valuable opportunity to inform a wide audience, to capture potential readers, and to develop or expand your interactive professional network. Whether others will pursue the reading of your complete text may depend entirely on their assessment of this tiny sample of your writing, including its style. Once again, emphasize problem and content, not fieldwork techniques.

THE COPY EDITOR AND OTHER MAIN TEXT ISSUES

Before we tackle the body of text, we should look at the role of the copy editor, a person you cannot avoid. This will probably be someone you have never heard of, will never meet in person, and will never hear of again. But for a few brief moments in the life of your manuscript, she (because all my copy editors have been female, I refer to them as "she") will hold what may seem life-and-death power over your "finished" manuscript. She will have seemingly unlimited power to inform you of grammatical conventions, references omitted or inserted without a corresponding citation, sentences that do not make sense, and perhaps even question whether you have all your facts in order. Although she cannot single-handedly shut down the entire publication operation, she may be the final obstacle you must overcome.

If you are very lucky, you may find your copy editor a dream to work with; if not, she may, for a while at least, become your Worst Nightmare Critic. She may laud your work and marvel at your skill with words and concepts, or you may have to justify almost everything you have written. If you cannot possibly work with the copy editor assigned to you, you may be able to arrange for an alternate, but most likely you will have to find a way to get along. Her self-perception will be that she is looking out for *everyone's* interests, but you may feel that your own interests are secondary, and you only wish she would go away. Still, you do want the manuscript to be right and correct, and if she does her job well, she will strengthen it, eliminate errors, catch grammatical mistakes, and save you embarrassment. My warning is only that the appearance and authority of the copy editor may come as a surprise to a first-time author, especially if to this point you have built a cordial relationship with the only person (your friendly acquisitions editor) whom you expected to shepherd your manuscript through to completion.

Even if you have worked productively with the copy editor, this will be your last chance to check *all* changes made to your manuscript. Copy editors work on your behalf, but they work *for* the publisher, and changes they deem improvements are incorporated directly. When copyediting was done on hard copy, changes were easy to detect. Now that most copyediting is done on disk, request that a copyedited disk on which changes *have been highlighted* be sent to you for final approval.

With the text of your study firmly in place, and the copy editor satisfied, some further decisions must be made as to where things go. Footnotes serve as a prime example, although, like several related decisions, this one may already have been made for you. There are also some things you can do to improve readability independent of the text, such as giving careful attention to headings and subheadings, paragraphing, and the judicious use of graphics.

My comments throughout this monograph deal with the preparation of printed text, not with alternative forms of textual representation such as ethnopoetics or performance texts such as ethnotheatre. "Performance texts" have narrators, drama, action, and shifting points of view with materials that are variously "turned into poems, scripts, short stories, and dramas that are read and performed before audiences" (Denzin 1997:91). Frankly, I'm too old, too traditional, too wedded to a text-dependent career to be a major contributor to such efforts.[2] Keep in mind that there will always be a lot of us traditional types around. I think it prudent for anyone tempted by new and innovative approaches to do the "printed text thing" first, before exploring less conventional alternatives for getting the message out. Further, as Amanda Coffey cautions, such textual practices expose the author to additional forms of critical scrutiny, not only to "getting it right" as a social researcher but also getting it right as a successful poet, playwright, or creative writer (Coffey 1999:152). Even a purely "textual" approach can be supplemented and enhanced to appeal to a wider range of readers.

Headings, Subheadings, Paragraphs. Unless you write seamless prose, take a final look at your use of headings and subheadings and at the length of your paragraphs. Short sentences and short paragraphs make for comfortable reading, although academic authors are not inclined to write that way. If you can find no other basis for dividing your long paragraphs into two or three shorter ones through efforts at editing, then be somewhat arbitrary about it. Give your readers a break by taking one yourself. No hard and fast rule, but as a guideline, try to have at least two or three paragraph breaks on each printed page.

Consider as well whether readability would be improved by the insertion of more headings and subheadings. When important points seem not to draw the attention they deserve, or when the text seems to jump abruptly from one subtopic to the next, the addition of a heading or subheading may help signal the transition. I mention these devices as writing "tricks" here because if you haven't been able to break up your text through relentless editing, then you must do it mechanically, with an eye toward achieving an aesthetic balance between space and text.

Footnotes and Endnotes. Notes that accompany text are properly considered text rather than back matter. Nevertheless, with some publishers, certain traditions, and most journals, endnotes *follow* the text rather than accompany it. That is, notes are placed at the end of each article or chapter (as I have done here), or at the back of authored books. You may have no choice as to where the notes will be placed, but you certainly have control over the number and quality of them.

One resolution to the dilemma of seeing the notes separated from text is to write without footnotes. All your references—to your own fieldnotes or interviews, to sources formally cited, to ideas gleaned from others, even your "editorial asides"—can be embedded in parentheses in the text where they appear. That may result in some long, unwieldy sentences, and you may find yourself "editing out" entries because they interrupt rather than enhance the text. Perhaps you will discover that you do not need footnotes after all.

If you find that you cannot dispense with footnotes, try to keep them to a minimum and exert what influence you can to keep them near the text that prompts them. If you are given no choice in these matters, it is because they are considered to be questions of format and style dictated by tradition and, in journal publication, by economies of time and money. It once cost more to keep footnotes on the page corresponding with text because lines of type had to be moved (literally, by hand) to accommodate them. Computerization has eliminated this as a technical problem, but old habits die hard, and journals and publishers can prove as hidebound as academic disciplines in leaving things as they were. Journal articles are not usually so lengthy that it is a burden to locate accompanying notes printed at the end of an article. With longer works, I find it exasperating to have to search for endnotes that have been collected along the way and deposited at the back of a book.

Only for fields like history and biography that traditionally thrive on citations to other sources, can I think of a rationale for separating notes from text. Even there, the consequence—and resulting paradox—is that scholars writing in these traditions are forever interrupting themselves, sometimes giving their footnoting such attention that it takes on a life of its own to comprise a study-within-a-study. Although well-established in the disciplines where it is practiced, excessive footnoting does not provide a good model for field-oriented researchers whose primary sources deserve primary billing. We may be chided by scholars in other disciplines for "making up our data," but that ought to work to advantage to the extent that our participation is always firsthand and genuine.

I write early drafts without footnotes. I allow myself considerable excess in using parenthetical comments within sentences, occasionally writing a parenthetical paragraph as well. During revision, I reexamine these parenthetical comments to see if I can incorporate them into the text. Any remaining tangents, explanations, and asides are reviewed critically, with an eye for turning them into footnotes. My earlier preference was to avoid footnotes entirely (the first edition had none). That was no doubt a legacy from the days when we worked with typewritten manuscripts and footnoting presented a typing nightmare. Computer programs now keep track of them for us.

The advantage of such notes is that they allow the main text to go forward without interruption. Some authors use footnotes effectively, and a few delight us with them, but I regard them as something of a habit-forming affectation in academic writing. Like underlining, or using quotation marks to set off "cute" words, as we used to do, or *italicizing* and **boldface** letters readily at our fingertips through word processing today, footnoting can lose its effectiveness through overuse. Footnotes themselves can be as disruptive as the nesting sets of parentheses that characterize the writing of some academics.

Authors who want to avoid the footnoting ritual yet make their sources and explanatory comments readily available have yet another option. Instead of indicating endnote references in the text, one can credit all quoted sources and provide additional comments in a final section devoted to endnotes that supplement the text. A brief repetition of text or an excerpt from quoted material is sufficient to facilitate identification.[3]

Charts, Diagrams, and Tables. Until you create a manuscript of your own, you may not be aware of the proper and distinct formats for tables, charts, diagrams, figures, maps, and so forth, or customary ways of representing statistical data. This is the kind of detail to which publishers (and graduate schools, if you are writing a thesis) pay close attention. You will undoubtedly be informed by a publisher that you are expected to provide "camera-ready" copy if your text is accompanied by figures or other artwork. Exactly what constitutes camera-ready material varies widely depending on the technological expertise and the extent that a publisher wishes to assist or to maintain control, so it is essential to confer directly in each case as to what is required and who is responsible for doing it.

When preparing a manuscript without a particular publisher in mind, you ought to be able to anticipate most of the conditions that will later be laid down. The time to get tabular material in order is when you prepare it initially. Formatting is part of it. Attention paid to detail as you proceed will pay off handsomely later when you can attend to new problems rather than have to circle back to redo or recheck everything you have already set in place.

There are important decisions regarding the level of information that accompanies such material. As already discussed, the obvious condition is that tables and figures must be accompanied by sufficient information that they stand alone; they should not be dependent on prose within the text to make sense. Careful attention to subheadings can help, but the choice of a clear, accurate, and adequate title for each table, chart, or diagram is the most important feature to attend to. Tables themselves can be accompanied by their own independently numbered footnotes to offer fuller explanation as, for example, why percent figures do not add to 100% or why *N*s vary from entry to entry.

A Note on Using Statistics in Qualitative Study. I was surprised to read that the use of simple descriptive statistics seems to have declined in recent years, at least among sociocultural anthropologists publishing in their own journals (Chibnik 1999). Whether statistics have a place in any *particular* study must, of course, be judged in terms of focus and purposes, but it may be time to remind ourselves to count or measure whatever warrants being counted and measured and to summarize and report statistically when appropriate. I do not recommend "throwing in a table or two" simply to make one's study appear more rigorous, but there are times when a great deal of data can be summarized in a table. There also are those more quantitatively oriented among our audiences who are consoled when they find such treatments. If you have a preference for uninterrupted prose, consider summarizing relevant data in tabular form presented as appendices (discussed below).

Artwork and Photography (including Cover Design). At several points, aesthetic and practical decisions (i.e., cost and feasibility, permissions) must be made about how a manuscript will look. These include big questions as to whether photographs or other artwork are to be included and stylistic questions about type size, font, even whether to use icons or drop-cap letters at the beginning of chapters. If you are publishing in a journal or an ongoing series, most decisions have already been made. With publication of a book, there are some one-of-a-kind decisions, at least as far as you, the author, are concerned—provided someone is thoughtful enough to ask your opinion. My experience is that as publication draws near, there are so many details to be taken care of that decisions about the artwork—the cover design in particular—are likely to be forestalled and then, at the last minute, rushed. The only power you may be able to exercise is to veto ideas that do not seem to work or that threaten the overall integrity of the production.

Authors often leave these matters to chance, as though choices like cover design or color are none of their business. I know of few academic authors genuinely delighted with the cover art for their books, and some have been quite dejected by choices over which they felt they had no control. Although you are unlikely to have the final say, it certainly won't hurt to make your preferences known. Would you rather have impressionistic art, a mechanical design, an abstraction, or something quite lifelike? Do you want the cover to reflect something of the contents of the book or simply to be aesthetically pleasing? You might send the publisher color photocopies of covers that appeal to you and explain what you like about them. In that way, you may influence the cover decision in spite of possibly being denied a "final say."[4]

I am also surprised at the number of people who have told me that the cover design influences whether they even bother to browse a book. In my opinion, the best cover design I have had—a cartoon-like sketch of two knights jousting, reflecting the combative mood described in *Teachers Versus Technocrats* (HFW 1977, reprinted 2003)—was drawn not by a cover artist but by a friend (Jerry Williams) who was the set designer for the university's theatre department. I told him what I envisioned; he was able to capture the idea in pen and ink.

Fieldwork often includes photography, and photos certainly can enhance text, just as poor reproductions, or photos only tangentially relevant to the text, may detract from the overall quality of the finished product. Don't succumb to allowing anything schlocky to diminish the text you have struggled to complete. If a picture is still worth a thousand words (pre-inflation), keep in mind that photos of poor quality can detract by at least the same amount. As a guideline, if you feel you ought to apologize for the quality of any of your photographs—even if they are the only ones you have—why not leave them out? Avoid having to apologize for *anything* in your work. Conversely, if your photographs are of superior quality, do more to feature them. Integrate them with the text, perhaps even select one (or form a collage of several) for the cover.

THE BACK MATTER

Appendixes and Supplements. Appendixes (or appendices, following the Latin) are auxiliary materials added at the back. Tables, charts, maps, and diagrams critical to the text are ordinarily integrated within it, but there may be additional material an author wishes to make available. A practice among qualitative researchers is to excerpt relatively brief portions from important sources—interviews, especially—in the text, augmented by fuller typescripts in an appendix. In that way, detailed information can be made available for the scholarly reader without burdening the text with lengthy transcripts. Similarly, interview schedules or questionnaires are sometimes included in an appendix. Such plans may be thwarted by a publisher on the grounds that the material is of limited interest and can be obtained through direct contact with the researcher. (We never quite dissuade ourselves of the belief that our readers are deeply interested in every last detail of our studies!)

Another use of an appendix is to provide additional illustrative material or case histories that supplement the main text without interrupting the account. When one's audience is presumed to be quantitatively oriented, yet the researcher feels that descriptive data provide critical information and perspective, an appendix can supply such information without requiring explanation or apology. My hope is that "closet" qualitative researchers aware of the potential contribution these approaches make, yet reluctant to go public on their behalf, can follow a progression in their work that increases the qualitative dimension in stages, a little at a time. Case histories or comparable descriptive material can be slipped unobtrusively into an appendix in one's earliest studies, subsequently to be given more prominence (e.g., incorporated into the main text), still later achieving chapter status, and eventually given center stage. The progression from a rigidly quantitative approach to an essentially qualitative one in the careers of some research luminaries should not go unremarked.

The terms *supplement* or *supplementary materials* can be used interchangeably with appendices. If the additional materials are so voluminous that they are bound separately, they are usually labeled as a supplement. A caution: The bulkier those appendices or supplementary materials become, the more you need to ask whether you still believe that data "speak for themselves." If data do speak for themselves, there ought to be a great demand for original field notes and full-length interview protocols. Is there?

References/Bibliography. The most important back matter in scholarly publishing, and of immense help to colleagues, is the section for references. To readers familiar with the literature in a particular field, an author's list of references provides a quick and fairly reliable guide to his or her disciplinary or professional orientation, as well as to the depth and currency of that orientation. When I need a quick gauge on researchers whose works are unfamiliar, I check their "quoting circle"—the authors and studies they cite.

Although the two labels, References and Bibliography, continue to be used somewhat interchangeably, a distinction between them has come to be widely recognized in this age of information—and publication—overload. Bibliographies retain their traditional definition as "lists of works" on a subject, the kind of comprehensive-but-focused guide to the literature prepared by a resource librarian or someone pursuing a highly specialized interest. There was a time when a scholar making a new contribution was expected to provide a comprehensive list of all previously published material in the field. Those lists were properly labeled Bibliographies.

Today such Renaissance-thoroughness is seldom seen and no longer ordinarily expected. It has become increasingly difficult to remain up-to-date and in command of the relevant literature, even in highly specialized subfields. Those who doggedly try to keep up with what everyone *else* is writing often have difficulty finding time for commensurate writing of their own.

As a result of this information explosion—in quantity, if not always in quality—bibliographic thoroughness expected in an earlier day has been replaced by expediency. Instead of compiling comprehensive bibliographies, one is now expected to provide references *only* to works specifically mentioned in the text, such citations to be collected under the label **References** or **References Cited**. The guidelines as to what constitutes a legitimate reference are quite explicit: If no citation appears in the text, an item cannot be included in the references. As one academic journal advises its contributors, if you can untangle the logic, "All entries in the reference list must be cited in the text and vice versa."

Inexperienced authors are often caught unaware, creating extra work for conscientious copy editors—and embarrassment for themselves—when informed that they have included, among their references, works they have not cited or, the complementary sin, have included citations in text for which no reference is provided. Inventorying such irregularities is one of the early and easier tasks for your copy editor.

One drawback of the current practice is that for any and every reference an author wants to include, a citation *must* appear somewhere in the text. Journal articles often contain a telltale sentence listing in perfunctory fashion all the ought-to-be-mentioned classics that, quite likely, will *not* be mentioned again. One way around such rigid citation practice is to combine the reference and bibliographic functions under a more flexible title such as **References and Select Bibliography** or **References and Further Reading**. A specialized topic might even warrant a separate list of "Recommended Readings." Another alternative, the only option for complying with most journal formats, is to review the classics in a footnote (or endnote) that lists important prior works, perhaps noting where one's intended contribution fits among them. Because citations appearing in footnotes are included among the references, the classics receive due recognition without the shoddy treatment sometimes apparent when they are simply listed in pro forma fashion within the body of the text.

Embedding critical citations in the text rather than in footnotes or endnotes not only reduces the need for footnoting but also weans us from the practice of employing Latin abbreviations unfamiliar to today's scholars. In place of *ibid., loc. cit.,* or *op. cit.,* when a citation is to a work previously cited, only a page reference is required. When the citation is to a different work, or there is any possible confusion, simply repeat the author's name and year of publication, along with specific reference to page numbers, as appropriate. (If we now could get authors to stop putting a period after *et* when they abbreviate the phrase *et alia* in reference to multiple authors, our ignorance of Latin would no longer be so apparent. In the meantime, some thoughtful programmer would do the academic world a favor by including *"et."* among the misspellings to be rounded up at spell-check time.)

Along these same lines, in the *first* citation to a multiple-authored work, *all* authors ordinarily are identified, even if it seems that every graduate student on the project—or doctor in the hospital—got in on the act. In subsequent reference to the same citation, *"et al."* is acceptable after naming only the first author. Some guidelines suggest rather arbitrarily that you needn't list all authors (except for the full citation that must appear in the references) if there are more than six. In any case, authorship shared among several contributors is not a common practice among qualitative researchers. It is a legacy from laboratory science in which "authorship" is shared among those who participate in the theoretical or experimental work as well as in the write-up.

In qualitative research, where the writing can make or break a study, I suggest that only the principal author(s) be identified. Minor collaborators, field assistants, or seminar members can be identified in the acknowledgments, where they do not confound citations or imply authorship. Given current interest in collaborative research, coupled with misgivings about the lack of adequate recognition collaborators have sometimes received in the past (particularly with life history accounts), today we find authorship more generously shared. The authors of "other people's stories" seem especially careful to acknowledge each party's role in what is regarded as authorial partnership.

As with many of the details reviewed in this chapter, you may have no choice about reference style once a study is accepted for publication. An editor may send you the publisher's style sheet; refer you to a recent publication, journal issue, or web page; or point you to whichever style manual (and edition) currently serves as the standard. Graduate schools typically expect dissertation writers to follow current practice in their academic discipline as exemplified by its leading journals.

Although you may have no choice in selecting the style for a particular publication, these are discretionary matters and you can take comfort in recognizing the wide range of practices extant. My personal choice for reference style is that of the *American Anthropologist,* which is the style followed in this book. I like it not only because of its obvious link with the ethnographic tradition but for several other features. It is "clean" (no quotation marks, parentheses, or underlines/italics unless they appear in the original); it is complete (no abbreviations; capitalization follows the original source; authors' full names may be used); and it is elegant (authors' names appear only once, on a separate line that precedes all references to their works, listed chronologically beginning with the earliest).

Working in an ethnographic tradition in the psychologically dominated field of educational research, I have often found it ironic to be directed to put my references into APA style (i.e., consistent with the current edition of the *Publication Manual of the American Psychological Association*),[5] especially when presenting or discussing research carried out in an ethnographic tradition. Admittedly, however, APA renders a service by offering a widely accepted standard for authors and editors alike, just as the Modern Language Association of America offers authors in the humanities with its *MLA Handbook for Writers of Research Papers*. Personally, I recommend that researchers pursuing ethnographically oriented fieldwork familiarize themselves with "AA style" and employ it if given that option. However, in the broad field of qualitative research, or in a discipline like sociology, that prior to 1996 recognized no one single style as standard, the best advice has always been to follow the style of a major journal in the field.

Glossary. The Latin root of this term refers to any difficult word requiring explanation. A glossary is an alphabetized list of such words. An author also may supply a list of abbreviations or foreign terms appearing in the text, accompanied by explanations or definitions. The question of whether readers might benefit from such help offers a final opportunity to reflect on one's intended audience(s). We too easily lose track of how specialized our research topics become or how much insider language we have adopted for our own. If you are writing only for your fellow insiders, there will be no need for such lists. But are you? Technical terms come first to mind, but postmodern authors might have found broader appeal had they recognized that not even their everyday language was shared by all.

Index. For years I managed to escape the alleged drudgery of having to prepare an index. When it finally became necessary to prepare one, I was surprised at how the task resembled organizing data of any kind. Having now created an index for each of my recent books, I have been pleased with the result and surprised at how helpful an index can be, even to the author! Indexes make our studies infinitely more useful as scholarly resources. I regret the lack of an index in anything I published prior to 1995; it is a serious shortcoming.

A publisher may insist that an index be provided. You can pay to have one done if you have neither time nor patience for compiling it yourself. Time is a real concern: Indexing cannot be completed until page proofs are in hand, and at that point, virtually everything else has been taken care of, so at the last minute, the author may be the one who holds up the works. No problem if you have anticipated this step and have prepared a working draft of an index. And no doubt the best person to prepare an index is the person who wrote the book. As William Germano points out, "No one but the author can do the index the way the author wants it" (2001:178). That is the person most keenly aware of a work's underlying purpose and concepts, not just someone plodding mechanically through the text concordance-like to identify key words.

Software programs designed to assist with indexing tend to pull one in the latter direction, because they deal more readily with key words than with concepts, although indexing is but one more example of how microcomputers can assist in compiling and organizing data from the ground up. However, my own author-compiled indexes were compiled from the top down. I was guided in each instance by the stated purposes and intended audience of the book and by the major ideas and concepts one might expect to be addressed in it. I also borrowed useful subject headings from the indexes of several books comparable to mine. From the borrowed lists and my own inventory of likely headings, I compiled a tentative draft of the new index. Then I began "testing" my developing index against the text, adding or consolidating needed categories, deleting redundant or unused ones, until I had a fairly comprehensive index. This, in turn, was fine-tuned in the process of being paginated against page proofs. Publishers may have their own guidelines as to format. The *Chicago Manual of Style* provides an authoritative general source.

I compiled a name index separately from the subject index for each book. The former was easy to do, and it served as a double check on whether I had included all the authors I had cited and vice versa. By titling it "Name Index" rather than Author Index, I could include *all* relevant individuals mentioned, not just those cited as authors. An opposite guideline prevails in preparing a Subject Index:

To warrant inclusion, topics should be *developed* within the text. A subject index can become as complete and detailed as its compiler chooses to make it. It can also be an unwitting victim of space restrictions, for it is the only section where the allotted number of pages can be decreased at the last minute to meet stringent page limits determined at printing.

I understand that authors who provide their own indexes tend to compile shorter ones with each new book. If that is true, is it because they get better at it or grow weary of the chore? Having indexed books of my own, I now fully appreciate and *expect* to find an index in academic books written by others.

You can "hire out" the indexing chore—any publisher can put you in touch with people who do indexing for a living—but I must warn that the process can be costly, especially if you are dissatisfied with the end product after spending the money. Don't give up too quickly on the idea of doing your own indexing—you are the only person who understands completely what you intended to cover, and the task is nowhere as difficult as we old-timers like to claim.

Page Proofs and Proofing. The rush I feel when I receive page proofs of a forthcoming work is at once literal, figurative, and ambivalent. After what always seems inordinate delay, an author cannot help but wonder at receiving either the copyedited manuscript or a set of printer's proofs by express carrier with instructions to correct and return everything within 72 hours! With page proofs, the rush is also sensory, often more exciting than seeing the finished product several weeks or months later.

The ambivalence stems from the realization that words that have been in flux and "setting up" for so long are about to become permanent. In this final pass, you are asked only to ensure that what will appear in print corresponds with your manuscript. You can catch the printer's follies, but you may have to live with your own. If you have been asked to provide a disk along with the hard copy of your manuscript, you may be surprised to discover that *you* are the source of little errors that make their way into the final copy. If you are able to negotiate even minor changes at this late stage, you also may be required to accompany your request with funds to cover additional costs. Most likely you will be instructed to do absolutely no rewriting. The lesson is straightforward and so is the moral: The time for editing is past.

Therefore, regardless of how weary you may have grown of your manuscript, how anxious you are to be done with it, force yourself to make one last read-through of the final draft before you send it off. Visualize your words *as though* they have appeared in a book or journal. This *is* the version that counts! It may not be too late to make critical changes, although you won't make any friends doing so at the last minute. Better to see yourself as gradually letting go, for in a sense the manuscript is no longer in your hands. Make sure you keep a backup copy of any last-minute changes or corrections, then send the page proofs back, ready for the printer. Your manuscript is about to begin a life of its own.

SUMMING UP: ENDGAME

There are many details to be "looked after" if you have prepared material that is going to be published. Some details are not optional, others may confront you with a choice, and it is to your advantage to have given them some thought beforehand. Keep your original purposes clearly in mind with every decision you make, and do not hesitate to make your preferences known.

• Carefully read and follow the guidelines for submitting manuscripts to a journal or publisher. You may be able to negotiate some requirements, but in general, you will be expected to know and to observe format specifications.

• Don't allow the seemingly small tasks associated with Finishing Up to be given short shrift as "last-minute details" attended to hurriedly. Your final title and table of contents, your bibliographic citations, your abstract, your indexes, anything you prepare may be judged as a sample of your scholarship and writing. Be sure they are accurate, informative, and well written.

• Don't overburden your account with prefaces, introductions, forewords, acknowledgments, dedications, etc. Get to the point. If you want to chat more informally with readers, consider doing so at the end of your study, when they are better able to decide whether they want to know more about you or your work.

• When making additions to your bibliography or inserting quoted material from informants or other sources, get the details in the format you need, get them right, and be done with it. Don't leave the chore of checking details to the last minute when it is so much easier to get things right the first time.

- Make sure that any supplementary material you add, such as photographs or charts and tables, are appropriate, of high quality, and instructively labeled.

NOTES

1. In that regard, I had better acknowledge George Spindler as the source of the idea introduced earlier of theory as "making work," lest he get after me again! We discussed some of the material presented in Chapter 4 during a visit in May 2000.

2. I did collaborate as dramaturg with Professor Johnny Saldaña, Department of Theatre, Arizona State University, on a scripted performance adapted from the Brad Trilogy (see HFW 1994, Ch. 3, 7, 11; also 2002) titled "Finding My Place." The 90-minute "performance text" was presented at the Advances in Qualitative Methods conference in Edmonton sponsored by the International Institute for Qualitative Methodology, University of Alberta, February 2001. (A transcript of the play is included in HFW 2002. For more on performance texts or performance ethnography, see Denzin 1997, Ch. 4; McCall 2000; Saldaña 2005.)

3. For an example, see *Culture: The Anthropologist's Account* (Kuper 1999), in which 247 pages of text are followed by 38 pages of explanatory notes and citations. Only one formal footnote appears in the body of the text, accompanied by the author's observation that footnoting itself is "a particular focus of deconstructionist analysis" (p. 214).

4. For further discussion about cover design, particularly in using photographs, see Kratz (1994).

5. You can find APA footnote style at http://www.apastyle.org.

CHAPTER SEVEN

Getting Published

As of today, here is almost every single thing I know about writing.

—*Anne Lamott,* Bird by Bird, *p. xxxi*

With Anne Lamott, I too can say that as of today, here is almost every single thing I know about writing. I conclude with a few thoughts about academic publishing. As I look back, my experience with editors and publishers seems to have involved a lot of luck. It is difficult to distill lessons that might serve others except for the need for perseverance. How publishing will change in the near future, I dare not predict, for I have had a difficult enough time keeping up with its already-changed nature and the constant restructuring of publishing houses. To my surprise, I discovered that my own publishing experience has involved some forty-four professional organizations, commercial publishers, and university presses.[1] The lesson is that each publishing opportunity will prove a unique experience, not only for you but for your editors and publisher as well. Make the most of it.

Be aware that I have addressed only academic publishing here, which is not like writing a book of poems or fictional prose for publication. My understanding is that publishing in almost any other form requires a completely different procedure. One usually begins with an agent—someone who represents you to a publisher; unless you have some special contact, finding an agent will be your first chore. In that sense, academic publishing is easier—you are your own agent. That is one difference. Another may be the amount of royalty you can expect. Unless you are an old hand at this (in which case, I repeat, you would not be reading this book), you will learn what colleagues mean when they tell you that writing a book is an act of love.

Of course, like everyone else in academia, I do have some ideas about how to get published with scholarly publication. One of the best ways to locate a suitable academic journal or publisher is to ask around among active and published researcher-colleagues in your field. Not those who profess that they *ought to be* writing and publishing, but those who actually are. You might also do some reading on this specific topic, such as Powell's informative case study approach to understanding scholarly publishing, *Getting Into Print* (1985), or Germano's more recent and highly readable *Getting It Published* (2001).[2] But prepare yourself: Regardless of the magnitude of your just-completed research—whether conducted pre- or post-PhD—it is not too likely that you will be any more successful in getting a book published academically than you would be with commercial publishing.

That is not to say you shouldn't try. There is always the possibility of connecting with the right publisher or editor at the right moment. But in addition to looking for ways to publish a monograph-length study, consider writing up smaller sections as journal articles. The fact that a more comprehensive version of your work exists elsewhere—if only in your dissertation or copies of a final report available from you—frees you from having to recount everything while trying to say something. If there is to be a "full" account, I suggest you draft it first, even if you doubt that you will be successful in securing a publisher for it. With a full account written, you can look for ways that shorter pieces can be developed from or about it. Pursue those assignments "one bird at a time." Don't announce how many shorter articles you hope eventually to produce; get busy and write *one* of them. "A bird in the hand. . . ."

If your writing has been done with an eye to promotion and tenure, be aware that journal publication is ordinarily much faster than publishing a book or monograph, and publishing in an electronic journal is even faster. In my experience, chapters invited for edited volumes, although a virtual shoe-in for publication, take the longest to reach the shelves; it takes only one laggard among contributors to jam the works. Should you be among the early contributors, it may be *your* contribution that seems most dated by the time the book appears.

Long delays in publishing, regardless of the cause, are never a good omen. New materials arrive on publishers' desks every day, new editors replace old ones. Manuscripts become less publishable the longer they sit, even when sitting in publishers' offices. Editorial promises get reinterpreted, forgotten, sometimes flat-out broken. The game of "musical chairs" that large publishing houses play as they worry about corporate "bottom lines" and get shuffled among mega-corporations has exacerbated the problem and left authors with little recourse, especially with qualitative studies that at best produce modest returns. We are expendable.

Horror stories abound of manuscripts that were never published. I would not pay them much heed. Never allow yourself the luxury of total despair. I once had an enthusiastic editor at a university press write that he was "interested" in publishing a book-length manuscript I submitted for publication in a monograph series. He lamented that at the moment he did not have sufficient funds, and with an estimated 250 printed pages, plus photographs, my manuscript would be "relatively expensive" to publish. I wrote him off, interpreting his comments as a gracious rejection, but a rejection, nonetheless.

I began exploring other possibilities. I responded eagerly to another editor's response that, if reconceptualized and shortened, my manuscript might fit into his new series. Unfortunately, the abridged and (too slightly, it seems) refocused study did not meet the second editor's expectations. I now had *two* monograph-length versions of my study, no publisher in sight, and a somewhat topical account that might quickly become dated. In spite of its general social interest, I felt I should not invest more time on the topic. Its focus, the result of fieldwork in southern Africa during a year of sabbatical leave, was tangential to my scholarly interests, except for the ethnographic experience itself (HFW 1974a).

Then, unexpectedly, a letter arrived from the first editor informing me that he had recently been allocated additional funds and was ready to put my manuscript into production. He wanted to know whether I had any last-minute changes. From the outset, he had every intention of publishing my study as soon as funds became available. That's exactly what his letter said, when I reread it more carefully.

Academic publishing houses, like academic journals, tend to carve their special niche, preferring depth to breadth. The publisher most likely to publish your qualitative study is already publishing qualitative studies. Publishers who already publish studies most like yours are most likely to be interested in yours as well, unless what they already have in print is *too* similar to what you have written (rather than closely parallel, and thus complementary), or recent marketing experience has made them skittish. There is no reason not to try to dissuade them on either account—nor is there any reason to think you will be successful in doing so. Carefully examine their publication list, describing how your work will augment their existing list of publications in print rather than dilute their market. Authors are not particularly attentive to publishers; publishers are attentive to those who are.

At professional meetings, invest time at the book exhibits and search out publishers interested in the topics and approaches that interest you. Do a bit of eavesdropping on conversations at book exhibitors' stalls. Granted, most visitors are looking for new materials and for studies to augment their teaching, or are simply trying to keep abreast of their field, if only by titles and authors. But broaden your gaze to include everyone at the scene, not just the consumers of research and the bright-eyed, bushy-tailed "publishers' reps" there to ring up sales. Lurking somewhere nearby (perhaps not at the booth; more likely off talking privately with other authors, but available to meet with you by appointment) are the acquisitions editors whose responsibility is literally to *solicit* manuscripts (i.e., discuss manuscripts and ideas with prospective authors) rather than to sell them. Their conversation is of an entirely different sort: They visit congenially with their "authors-in-print," talk to authors with manuscripts (or ideas) for getting into print, and occasionally propose topics to prospective authors along lines the publisher feels might be productive.

Too good to be true? Well, recall from the preface my story behind the origin of this monograph. Editor Mitch Allen took me aside while I was perusing the books he was exhibiting at a professional meeting, to tell me about a manuscript he wanted for the SAGE Qualitative Research Methods Series. The monograph he had in mind dealt with the subject of writing up such research. And he wanted *me* to write it! It happens.

If there is any selling done in such circumstances, authors with ideas are doing it. But listen to those conversations and you will realize how astute most editors are, how knowledgeable they have become about what is being written in *your* field. They may dampen your enthusiasm or redirect you to a competitor with that great idea you were certain they would covet for themselves, but most editors have a breadth of vision that can become an author's valuable resource.

To remain in business, academic publishers keep a close eye on their markets. Fortunately for us, that includes the library market and crossover tradebook market as well as the market for large-scale text adoptions. Still, it can be disappointing to realize that a fine piece of existing research, exquisitely written, may not be a candidate for publication because it will not "sell"—i.e., is not expected to command enough market to make publication feasible. What sells cannot be the only basis for conducting research, however. And clearly it is not. The market for qualitative studies is still comparatively thin, oversaturated by our own successful efforts to convince publishers of potential markets that never quite materialize. In their view—and experience—as few as one out of ten books is likely to make money. My guess is that within each subfield, a few studies—our own modest shelf of "classics"—account for most sales and are the studies to which we all point as precedent. As Mitch Allen once observed about this unique market, "The writers of qualitative research are the buyers of qualitative research. It is a closed system."

We might appear to be advantaged by having university presses as another publishing option. Ostensibly, their mission is to advance scholarship, rather than realize profits. But the day seems to have passed when university presses were willing to take chances by publishing materials with uncertain or thin markets. Increasingly, those presses not only have become self-supporting but are expected to make a return on the university's investment in them. Rather than serving as a fallback to guarantee that studies with thin markets will be published, university presses today tend to seek the better stuff, which they further enhance with their imprimatur. Although in an earlier day university presses were instrumental in making publication possible in esoteric fields, certainly including ethnography, today only a few of them devote much attention to publishing qualitative/descriptive work.[3] If you would like to publish with a university press, look for a match between your manuscript and the books already on their shelves rather than attempt a sweeping canvass to see if anyone is interested. Commercial publishers are even more leery of publishing research monographs. No reason not to try, but be aware how difficult it may be to find an interested publisher, especially for your first publication.

You may be better off to locate an appropriate *series* in which to publish rather than hope to publish your work as a separate, stand-alone piece. An alternative is to find a small publishing house able to minimize risks by minimizing costs. In your eagerness to get published, don't lose sight of the fact that small publishing houses also have small budgets for advertising: One can publish *and* perish, getting material into print that remains unknown. That is also the catch to the ease of desktop publishing. Being able to "publish" our own studies does not resolve the question of distribution, even if recovering out-of-pocket costs is not a major concern. So-called vanity publishing (publishing commercially at your own expense), or making your work available electronically, raises questions of legitimacy because it is often done in the absence of peer review. In the latter instance, the "legitimacy" issue joins a host of other problems as yet unresolved (such as ownership, copyrights, review, availability, etc.), although the opportunity to make your work readily accessible through electronic publishing is unprecedented.[4]

If you are successful in finding a publisher, your happy anticipation may give way to frustration as you begin to wonder whether your publication is one of the world's best-kept secrets. Rather than bemoaning how little your publisher seems to be doing to promote your study, take responsibility to help spread the word. Send letters or announcements to your professional colleagues. Advise your publisher of the journals to which your book should be sent for review (supply addresses and the name of the current book review editor, not just the name of the journal), and follow up independently to make sure that the material was received. You may be able to badger your publisher into distributing some complimentary copies if you supply names, addresses, and a rationale for your selection of recipients.

The real key to the marketing problem lies beyond the scope of qualitative research. Our studies are not adopted for classroom use on the scale that makes textbook publishing lucrative because qualitative studies are not easy for instructors to use. They are neither self-teaching nor self-evident. They *can* make teaching more exciting; they *definitely* make it more challenging. The best way we have to expand the market for qualitative/descriptive studies is by demonstrating their effective use in our own teaching. But that is another story, maybe even another monograph: *Teaching (Teaching Up?) Qualitative Research.* I have had my say in the matter (HFW 1994: Ch. 12, 13). Our responsibility as author/researchers is to make sure that, when sought, the studies are there, well researched and well written.

ALTERNATIVE WAYS TO GET IN PRINT

Although I "think writing" from the outset of a study and I begin thinking about a working title and Table of Contents almost as soon as I begin a research project, I do not think "publication" with that same single-mindedness. In our work, the research act is not really finished until our studies are accessible to others. *There is no such thing as unreported research.* The customary form for that documentation is a written account. When we set out to find a publisher for a qualitative study, we ordinarily have a completed project in hand, not just an idea or prospectus. We do not approach publishers with the hope that they will fund research, although they may be willing to underwrite some costs in manuscript preparation.

My preference with everything I have written is to do the writing *first,* then negotiate a contract. When my writing proceeds on the basis of an invitation or verbal agreement, so much the better, but I am uneasy knowing that someone already owns material I have yet to write.

Peter Woods, who offers excellent advice in *Successful Writing for Qualitative Researchers* (1999), takes exception to my "write first" approach. He calls it "risky business" and suggests seeking a publisher from the outset. You must decide for yourself whether there are compelling reasons for writing up what you have to say, market or no. My feeling is that when you have something to say, you should write it. Get the account recorded in your own words before submitting (literally and figuratively) to what someone in authority says you will have to do to get published. If writing is "in your blood," you know what I mean.

THE REVIEW PROCESS

Whatever eventually compels you to inquire about getting into print, I advise against sending anything longer than a journal article to a publisher or editor without prior communication and an explicit request for more. If it is "common knowledge" that unsolicited manuscripts have only a slim chance of being published (Powell 1985:89), then the secret is to get a manuscript solicited. To receive such an invitation, send the title page and Table of Contents, accompanied by a carefully prepared letter (addressed to a particular individual by name, if possible) explaining not only what you have written about but *why you have chosen that particular publisher.* Better still, should you be so lucky, have a colleague with contacts at the press telephone or write on your behalf, especially if you are exploring possibilities for publishing your (revised) dissertation.

Describe the current status of your manuscript and any unusual circumstances surrounding it, such as how soon you could send a polished draft, or problems with clearances, permissions, or any "conditions" surrounding publication. A broad inquiry might be sent to several publishers, but once you receive an indication of serious interest, stop playing the field. If you are tempted to browbeat publishers or journal editors by claiming that a manuscript is under consideration elsewhere, recognize how easily the ploy can backfire. Who wants to invest time and money in a go/no-go decision on a manuscript that may have already gone?

Journal publication seems the more realistic option for getting into print if you can pare an article down or "chunk out" something from a longer manuscript. Journal-length manuscripts circulate easily among colleagues and editors. You do not need an invitation to submit a manuscript to a journal. However, you might communicate with their editorial office prior to submission if you are uncertain whether the content fits within a journal's scope, or if an article presents some unusual problem such as requiring special graphics or exceeding customary length. Even in these cases, expect the reply, "Send it along and let us have a look at it." (Hint: I *always* find some excuse for checking first with the editor. Voilà! Back come the words I want to hear, "Send it along and let us have a look at it.")

An author needs to select journals with care and to demonstrate awareness not only as to their scope but to their formal requirements for submission. A cover letter that provides a brief introduction should also explain any aberration between the submission as made and the stated requirements of the journal. Minor deviations should be acknowledged and explained—for example, recognition that the citation style differs from that of the journal and will be reformatted if the manuscript is accepted. A manuscript should not be accompanied by an apology or a sweeping promise to do *anything* to get it accepted. Comply with requirements and let the manuscript speak for itself, as it will have to do when published. Remember that most editors of professional journals are themselves busy researchers and teachers who must get on with their own work. They have every right to expect manuscripts to be in polished and complete form and in what you assume to be the final version—even if they subsequently ask for revision.

Nobody relishes rejection. Having a manuscript rejected is always disappointing, to old-timers as well as new authors. The most difficult rejections are those that arrive without explanation or comment. Yet I know from having to pen such letters as an editor, and having to make sure they would not be misinterpreted as giving false hope or phony encouragement, that sometimes there isn't anything to say except "Thank you for considering this journal."

In spite of efforts at multiple, external, and sometimes blind review, the review process can seem capricious. One problem is that final editorial decisions do not, and cannot, rest solely on the basis of outside reviewers' recommendations. Accordingly, a rejection or two should not lead to a premature conclusion that a manuscript is unworthy. Pay close attention to specific suggestions or criticisms noted as shortcomings. It may be a good idea to share a rejection letter with a close colleague; subtle messages "between the lines" sometimes escape sensitive authors. And editor C. Deborah Laughton advised, never, ever interpret as rejection a letter that suggests only that you "revise and resubmit." Just do as directed.

Waiting for review provides another respite that can be turned to advantage. With the passing of time, one can usually return to a manuscript with a fresh look. If you are really brave, when your manuscript is returned, read it afresh as though you are a reviewer rather than the author. Should the review process become too protracted, you might even develop a separate working paper or oral presentation in which you describe the kinds of feedback you have received and how you are coping with the waiting period. I was having trouble "getting past" reviewers with one conceptual paper I was developing. I prepared a companion piece subtitled, "Notes on a Working Paper," in which I could take on my critics and give adequate attention to some excellent points that had been raised. My "Notes" made for a lively keynote address and subsequently were published in a conference proceedings (HFW 1991). Instead of beating me down, I treated the review *process* itself as a new source of data.

I never read anything of my own in draft—no matter how long I have been working on it—without a pencil or mouse in hand, alert for sentences that can be shortened, ideas that can be expressed more clearly, interpretations that can be strengthened. But once a manuscript is in production, compulsive editing must come to a halt. When something I have written finally appears in print, I read with whatever sense of accomplishment seems warranted, never with a sense of disappointment. Those are my words, my sentences, my ideas. (And, after all that work, they better be mine, just as I wrote them, unless I have been advised of any but the most minor of editorial changes.) I stand by them. At the time they were written, they represented my best effort.

ON NOT GETTING PUBLISHED

What if you are unable to publish your full account? And what if, realizing that its appeal is limited, you draft a couple of shorter articles about specific aspects of the work, but you are unable to find a suitable journal interested in publishing them? Is that the end of the world? Or your career?

Well, not getting published may not do much for your career, but after spending more than four decades in a university setting, I can report that I have never heard of an academic promotion or tenure decision based *solely* on someone's publication record, even if a decision is made on someone whose writing seems to have stalled. Failure to publish "enough" seems only a convenient peg on which to hang negative decisions. If you had access to the publication record of everyone promoted at most institutions of higher learning, I think you would be shocked at how little some people have published. (Presumably they contribute to their institutions in other ways. If everyone were busy writing, who would prepare the institutional reports, call all those meetings, coach the performers, or ration the office space and travel money?)

True, at so-called research universities, you must write, create, produce, or manage *something,* but it strikes me as unlikely that anyone whose motivation for publishing stems only from a preoccupation with tenure or promotion would turn to so time-consuming an activity as qualitative research. Such individuals should not be looking for alternative forms *of* research, but for alternatives *to* research that satisfy criteria for achievement and recognition.

There are numerous alternatives through which respectable contributions can be made to scholarship: developing synthesis papers or position papers, preparing annotated bibliographies, convening conferences. No doubt some, perhaps most, of your colleagues are publishing *something,* but take a critical look at how many are publishing **original research**. What you are reading here, for example, is experience-based and in a scholarly tradition; it deals with the sacred *topic* of research. But it is not research. Despite such humble origins, I expect it to find a place *somewhere* in the Great World Series in the Sky where academic achievement is recorded.

Assuming that you are committed to qualitative research, I urge you forevermore to regard writing as a vital aspect of the research process, rather than as an activity inexorably linked with publishing. Whether you publish is in no way as critical to your role as a qualitative researcher as whether you complete your studies by making them accessible. Every research effort must finally come to rest in some tangible, processed form. Unpublished field notes are not enough. Comprehensive field reports drawn from them, completed but unpublished papers, papers modestly reproduced under the aegis of your agency or department, papers or poster sessions presented at conferences, reports of your work available electronically—all these contribute and count toward scholarship and, as well, toward your credibility (and visibility) as a researcher who carries work to completion.

A successful career doesn't mean that every effort along the way has been a success. Even success can impose a formidable barrier to further writing, especially if it comes early in one's careers. Those of us who have been at this awhile hear whispers that we no longer seem to write "as well" or "as engagingly" as we once did. Nor does every publication receive the recognition we might feel it deserves. We, too, have batting averages. Nobody scores a hit every time.

And thank goodness they don't. There's so much in print already. Not all our work needs to be published, certainly not in the slick format of expensive journals and books. For the most part, our purposes can be accomplished with less formal, less expensive formats, such as in seminar papers modestly circulated among colleagues *without* the awkward accompanying question, "Where should I send this?"

The good news is that if you are determined enough, you surely can get published someday, somewhere, on screen if not on paper. Electronic publishing is opening things up in ways previously unheard of. Such journals already fill an important gap, offering a quick and inexpensive way to "publish" by putting articles directly on the Web without having first appeared in print. The once-tedious process of collegial review has also been speeded up through rapid communication among authors, editors, and reviewers, with possibilities being explored that should help the review process become both more rapid and more widely shared.

In time, electronic publishing may even become an entry-level *prerequisite* to publication in printed journals, thus allowing for better screening and selectivity in what is formally published. That might help ensure that more of the better stuff makes it into print, and that what makes it into print includes more of the better stuff, without unnecessarily cutting down on opportunities for everyone to make their work accessible. But electronic journals are variously perceived as alternatives *to* publishing as well as alternatives *for* it, and we have yet to see how, where (i.e., in which fields), or whether they will come to be regarded vis-à-vis "traditional" publication. Neither my eyes nor my patience are well adapted for reading large bodies of text on screen, but I have been dragged into the computer age nonetheless. We can brace ourselves for extended discussion about electronic publication, as the pros and cons are debated about something that is already happening.

By all means, stay with any worthwhile study until you have seen it through to the completion of a clearly conceptualized and well-written account. Make sure that, in some form, accounts of your research reach the hands of the people who share your interests. Without insisting that you "must" get published, ask their advice in helping to assess the audience you should reach and how much additional effort on your part seems needed and warranted. Published or not, you've written up your qualitative research. Your work wasn't completed until you did. Or, if not actually published, at least it is a beginning, and that is something!

FINAL THOUGHTS

- No one other than you yourself need ever see your early drafts.

- Until you have a rough draft of what you have to report, there is no chance of improving it. Start there.

- Solicit the views of editors and publishers about the topics or ideas you have for publishing.

- There are several alternatives to publishing, ways through which you can make your (unpublished) research available to others.

- Keep always in mind: There is no such thing as *unreported* research.

NOTES

1. Journal articles and chapters invited for edited volumes account for the wide variety. Looking only at authored books tells a different story. Two of my original books were published by university presses; two were solicited by my mentor, George Spindler; and the six most recent ones were solicited and edited by Mitch Allen. In more ways than one, both George Spindler and Mitch Allen have had a great influence on my views, experience, and opportunities for writing and publishing. Looking back, I recall that getting my earliest writing accepted was a slow process, so be very, very patient with yourself.

2. See monographs in the SAGE series *Survival Skills for Scholars* that deal specifically with publishing books (Smedley and Allen 1993) or journal articles (Thyer 1994).

3. For the previous edition, I thought it might be helpful to identify a few presses by name. I compiled a preliminary list and sent a one-page questionnaire. One press responded with their form letter of rejection. Another saw little point in compiling such a list. Amber Wilson at UC Press was kind enough to list 27 categories of qualitative study which the press was publishing at that time: African studies, anthropology, art history, Asian studies, classical studies, cultural studies, European history, film, fine arts, gender studies, geography, Jewish studies, Latin American studies, law, linguistics, literary studies, Middle Eastern studies, music, natural history, philosophy, photography, political science, religious studies, science, sociology, and women's studies (Source: personal communication 15 August 2000). However, six months passed before I received a response from *any* of the presses contacted. Fair warning: University presses do not seem to be bending over backward to solicit manuscripts.

4. The most recent review that I was asked to write (HFW 2008b) was published electronically. When *Teachers College Record* invited me to write it, I was not sufficiently attentive to the fact that their reviews appear only in electronic form. I accepted the assignment and soon received a copy of the book. By the end of the following month, my review had been

accepted and "published," but from my perspective, it simply vanished. I had not realized that the reading of reviews in *Teachers College Record* is reserved for subscribers and others who pay an annual fee for access to reviews. I am not a subscriber. Although they did offer me a PDF of my review, I cannot otherwise access it.

Appendix

Applications

I have tried to avoid the authorial voice or tone of the textbook writer that might make this monograph read like a classroom text. I am impatient with authors who conclude chapters with questions for me to "think about" or assignments for me to complete. But it may be helpful to suggest ways to make your reading and writing more interactive. If so, I can oblige with some thoughts and practical exercises that may assist you with your writing. These ideas may also suggest some ways for instructors to help their students become better writers.

Let me remind you of my attempt at the end of Chapter 1 (and again at the end of Chapter 2) to get you involved. I suggested that if you aren't presently engaged in writing up research, you might think about a project that conceivably you *could* undertake in the near future. For that hypothetical research, I suggested that you draft a problem statement (in writing) to focus your thoughts as you read further. If you do not have either a real problem statement or a hypothetical one in mind, let me suggest once more that you do that exercise before you turn to the ideas discussed below.

1. Try your hand at cutting through the thick. Copy a paragraph of particularly thick prose from a journal article or book chapter in your field. Try editing it, not in the sense of totally rewriting but to help the author say more clearly whatever he or she is trying to say. Then take a shot at editing a paragraph of something you have written. You may discover that donning an editor's hat instead of your customary author's one helps you to achieve the needed distance from the words in front of you.

2. Find an article written in the objective "third person" and rewrite a couple of paragraphs in first person. In the instance you have selected, does changing the "voice" from third person to first improve the tone? What else should you take into consideration before deciding whether to use first- or third-person language? In similar fashion, find an article written in the passive voice and see how it reads after you transform part of it into the active voice.

3. Ask someone—better yet, ask several people—to read aloud a selection you have written. Notice how the phrasing and intonation change with different readers. When you edit your material, you need to be able to "hear" how your sentences may be read—and misread.

4. Play with titles. Brainstorm possible titles for your current or next research project. Then list them in order, beginning with the most suitable. Are you able to discern a pattern among the titles to help you identify words or ideas that you want to be sure to convey? Does each proposed title convey sufficient information to catch the attention of possible readers? And what would be the likely categories in which your work would be cataloged by someone compiling an electronic bibliography from titles alone? How "cute" do you want your titles to be, and what is the corresponding risk?

5. Pushing the idea introduced in Chapter 2 of the mix among description, analysis, and interpretation a bit further, what might be the circumstances appropriate to make any *one* of these elements the primary focus of a project report? Can you think of any circumstance when each of the three elements might warrant equal attention? And can you anticipate how the way you apportion your attention among those same three elements is likely to shift during the course of a research career?

6. If you are presently working on a document, prepare a brief style sheet to accompany it. Put it in the form of a memo to the copy editor to raise questions about, or state your preferences as to, your handling of terms (e.g., *participant observer* vs. *participant-observer*), use of commas and the serial comma, preferred style for footnotes/endnotes, and any other questions that need to be resolved to give the document a consistent format.

7. As a resource, start a personal notebook of wise and pithy sayings, useful definitions, timely proverbs, theoretical insights, etc. I have labeled my notebook *Quotes*. What will you call yours? Don't

you already have some stuff to put in it? And where are you going to keep it? My choice is to keep quotable quotes in a bound notebook on my desk, rather than in a computer file; I want to be able to access it anytime. How will you use and keep yours?

8. I was once asked to provide a brief synopsis of an opera for a friend unfamiliar with it. I decided to begin my account by relating how things would stand as the final curtain fell. Then I kept reaching back into the story for the critical events that would eventually lead to that (presumably tragic) ending. Afterward, I wondered why we never see research reported that way. Think about presenting a research account beginning with how things presently stand, and what we have learned, and then drawing on relevant detail as needed to explain how we arrived where we did.

9. Sketch a "tree" of your own to show the relationships as you perceive them among the strategies that qualitative researchers can take. Feel free to use my "tree" as your starting place. Or, reconceptualize the whole qualitative research field in some other way, such as with a Venn diagram. Are you comfortable with the idea presented in Chapter 4 that participant observation serves as the *core* research activity in all qualitative inquiry?

10. In your own field, can you identify researchers who have made a major contribution to theory? Who are they, and what is the nature of their contribution? Do you have to go outside your field to find its prominent theorists? Bear in mind that most researchers are ready and willing to voice opinions *about* theory and the various roles it can or might play, but that should not be confused with theory-building itself.

11. From a popular magazine or professional journal, select an article that you feel would benefit from editing. Assume that it has not yet appeared in print but has been returned to a colleague with instructions to cut it "by about one third." Can you suggest where and how the author might cut to meet such stringent space limitations? Will lopping a word or two from each sentence reduce the overall length sufficiently?

12. Prepare a carefully worded abstract of an article or report you have written. (If you will be writing a dissertation, draft the abstract required for *Dissertation Abstracts*.) Stay within the prescribed word limit for journals in your field. Identify key words by

which the material can be indexed. How can you be sure the words you select will bring your work to the attention of the audience(s) you wish to reach?

13. Is there a research topic on which you are so knowledgeable that you could "almost" write up a study before conducting field-work? Try drafting it. Were you actually to pursue the topic after preparing such a draft, do you think your efforts would lead to a better designed research project?

14. In good fun, try to outdo the most serious offenders in your field by writing a parody of a pompous opening paragraph for a pro-posed study of no importance. If you are a member of a seminar, per-haps you can hold a competition to see whose paragraph is the most outrageous. Then solemnly resolve never, never to write such drivel or allow yourself to become involved with trivial pursuits in the first place.

15. (Extra credit for this one.) I keep urging you to get right to the problem statement, "The purpose of this study is. . . ." Still, it does seem a shame to have you start right off with a passive con-struction. Can you find ways to rewrite the problem statement so that you begin your writing in more lively fashion?

16. For your real or an imagined study, set up a "hanging file," assigning one folder for each chapter (if it is to culminate in a book) or section (if an article). You could do this on your computer or you can probably find an inexpensive hanging file in an office supply store or campus bookstore. A set of manila folders will serve as well. Into each folder, put a brief statement describing the likely contents of that section or chapter. Hey, if you can go that far, why not develop a rough draft of the material to go in each folder? If you can do that, you're well on your way to your next publication.

17. Begin keeping your own list of the possible publishers of the kind of study you are doing or intend to do. Get the names of contact persons there, and keep track of where you get your infor-mation. You might also want to keep a list of publishers that you are certain would not be interested in your work.

18. A final suggestion, this one for the instructor:

As a teacher, I wanted to hear from all my students on a regular basis, but there was no practical way they could all contribute equally in class discussions. So I needed to "hear" from them in writing. And often. Yet I cringed at the thought of reading and reacting to so many papers from each class each term.

I hit upon the idea of setting a one-page limit on all papers submitted during the course (except the term paper submitted in lieu of a final exam). For most students, that imposed limit was interpreted to mean all the words they could squeeze onto one side of a standard typewriter page. The papers did not have to leave any margins, and all references, as well as the writer's own name, were put on the back (the latter so that I would not know the identity of the writer, at least early in the term). In pre-word-processing days, the papers often had to be retyped two or three times before they "fit." Each retyping presented another opportunity for editing. With the arrival of word processing, it became easier to reduce the text to one page, and I needed a new rule about the smallest allowable font size to forestall students tempted to meet the space limit that way.

The unexpected dividend of the assignment was that it fostered disciplined writing, necessitated getting right to the point, and demanded careful editing. Although those one-page papers proved a torturous exercise for many, especially at the beginning of each term, the assignment also proved a wonderful way to help students become better writers. They had numerous opportunities to write, yet each assignment was of manageable length. And I could "hear" from each student often and on a recurring basis.

Acknowledgments

With two prior editions behind it, *Writing Up Qualitative Research* has benefited from the help of many people. I have not listed those earlier reviewers again; their names are included with each respective edition. For this revision, I now add my thanks to the panel of reviewers for SAGE who each offered useful suggestions: Judy K. C. Bentley, SUNY, Cortland; Pat Maslin Ostrowski, Florida Atlantic University; Joseph Wronka, Springfield College; and Sonja Peterson-Lewis, Temple University; and to an editorial team at SAGE consisting of Acquisitions Editor Vicki Knight; Project Editor Astrid Virding, who coordinated the effort; and my "dream" copy editor, Liann Lech. I am also indebted to Edith King, who forwarded suggestions for this revision from students in her graduate seminar at the University of Denver, to Katy Lenn at the University of Oregon Library for assistance in tracking references, and to Ken Loge, who gave help when I was at the point of throwing up my hands (or just throwing up?) in frustration at what my computer is capable of doing that I cannot. That marvel of technology seems to hide as many tricks as it reveals. Or so it seems for someone who did not grow up in the "computer age" but has had to grapple with it long after having mastered a number of routines now considered archaic. May you never suffer the frustration of correcting an error when typing an original and two carbon copies.

References and Select Bibliography

Agar, Michael H.

    1996   The Professional Stranger: An Informal Introduction to Ethnography. 2nd edition. New York: Academic Press.

Barrett, Paul H., Peter Gautrey, Sandra Herbert, David Kohn, and Sydney Smith, eds.

    1987   Charles Darwin's Notebooks, 1836–1844. Ithaca, NY: Cornell University Press.

Bateson, Gregory

    1936   Naven. Stanford, CA: Stanford University Press.

Becker, Howard S.

    1986   Writing for Social Scientists: How to Start and Finish Your Thesis, Book or Article. Chicago: University of Chicago Press.

    1998   Tricks of the Trade: How to Think About Your Research While You're Doing It. Chicago: University of Chicago Press.

    2007   Telling About Society. Chicago: University of Chicago Press.

Beer, C. G.

    1973   A View of Birds. In Minnesota Symposia of Child Psychology, Vol. 7. Anne Pick, ed. Pp. 47–53. Minneapolis: University of Minnesota Press.

Behar, Ruth

    1992   Translated Woman: Crossing the Border with Esperanza's Story. Berkeley: University of California Press.

Benedict, Ruth

    1946   The Chrysanthemum and the Sword: Patterns of Japanese Culture. Boston: Houghton Mifflin.

Bernard, H. Russell

    1994   Research Methods in Anthropology: Qualitative and Quantitative Approaches. 2nd edition. Thousand Oaks, CA: Sage.

    2000   Social Research Methods: Qualitative and Quantitative Approaches. Thousand Oaks, CA: Sage.

    2006   Research Methods in Anthropology: Qualitative and Quantitative Approaches. 4th edition. Lanham, MD: AltaMira Press.

Biklen, Sari K., and Ronnie Casella

    2007   A Practical Guide to the Qualitative Dissertation. New York: Teachers College Press.

Boellstorff, Tom

    2008   How to Get an Article Accepted at American Anthropologist (or Anywhere). American Anthropologist 110 (3):281–283.

Brettell, Caroline B., ed.

    1993   When They Read What We Write: The Politics of Ethnography. Westport, CT: Bergin and Garvey.

Bruner, Edward M.

    1986   Ethnography as Experience. *In* The Anthropology of Experience. Victor W. Turner and Edward M. Bruner, eds. Pp. 139–155. Urbana: University of Illinois Press.

Burke, Kenneth

    1935   Permanence and Change. New York: New Republic.

Campbell-Jones, Suzanne

    1978   In Habit: A Study of Working Nuns. New York: Pantheon Books.

Caulley, Darrel N.

    2008   Making Qualitative Research Reports Less Boring: The Techniques of Writing Creative Nonfiction. Qualitative Inquiry 14:424–449.

Cavan, Sheri

1966   Liquor License: An Ethnography of Bar Behavior. Chicago: Aldine.

Cheney, T. A. R.

2001   Writing Creative Nonfiction: Fiction Techniques for Crafting Great Nonfiction. Berkeley, CA: Ten Speed Press.

Chibnik, Michael

1999   Quantification and Statistics in Six Anthropology Journals. Field Methods 11:146–157.

Chiseri-Strater, Elizabeth, and Bonnie Stone Sunstein

1997   Field Working: Reading and Writing Research. Upper Saddle River, NJ: Prentice Hall.

Clifford, James

1988   The Predicament of Culture. Cambridge, MA: Harvard University Press.

1997   Spatial Practices: Fieldwork, Travel, and the Disciplining of Anthropology. *In* Anthropological Locations. Akhil Gupta and James Ferguson, eds. Pp. 185–222. Berkeley: University of California Press.

Clinton, Charles A.

1975   The Anthropologist as Hired Hand. Human Organization 34(2):197–204.

1976   On Bargaining with the Devil: Contract Ethnography and Accountability in Fieldwork. [Council on] Anthropology and Education Quarterly 7(2):25–28.

Coffey, Amanda

1999   The Ethnographic Self: Fieldwork and the Representation of Identity. Thousand Oaks, CA: Sage.

Coser, Lewis

1975   Two Methods in Search of a Substance. American Sociological Review 40(6):691–700.

Crapanzano, Vincent

1980   Tuhami: Portrait of a Moroccan. Chicago: University of Chicago Press.

Creswell, John W.

  2003   Research Design: Qualitative and Quantitative and Mixed Methods Approaches. Thousand Oaks, CA: Sage.

  2007   Qualitative Inquiry and Research Design: Choosing Among Five Approaches. Thousand Oaks, CA: Sage.

Deegan, Mary Jo

  2001   The Chicago School of Ethnography. *In* Handbook of Ethnography. Paul Atkinson, Amanda Coffey, Sara Delamont, John Lofland, and Lyn Lofland, eds. Pp. 11–25. Thousand Oaks, CA: Sage.

Denzin, Norman K.

  1997   Interpretive Ethnography: Ethnographic Practices for the 21st Century. Thousand Oaks, CA: Sage.

Denzin, Norman K., and Yvonna Lincoln, eds.

  2000   Handbook of Qualitative Research. 2nd edition. Thousand Oaks, CA: Sage.

  2005   The SAGE Handbook of Qualitative Research. 3rd ed. Thousand Oaks, CA: Sage.

Elbow, Peter

  1981   Writing with Power: Techniques for Mastering the Writing Process. New York: Oxford University Press.

Emerson, Robert M., and Melvin Pollner

  1988   On the Uses of Members' Responses to Researchers' Accounts. Human Organization 47(3):189–198.

Fetterman, David

  1989   Ethnographer as Rhetorician: Multiple Audiences Reflect Multiple Realities. Practicing Anthropology 11(2):2, 17–18.

Feyerabend, Paul

  1988   Against Method. Rev. edition. New York: Verso.

Fink, Arlene

  2005   Conducting Research Literature Reviews: For the Internet and Paper. 2nd edition. Thousand Oaks, CA: Sage.

Fitzsimmons, Stephen J.

    1975   The Anthropologist in a Strange Land. Human Organization 34(2):183–196.

Flemons, Douglas

    1998   Writing Between the Lines: Composition in the Social Sciences. New York: W. W. Norton.

Flinders, David J., and Geoffrey E. Mills, eds.

    1993   Theory and Concepts in Qualitative Research: Perspectives from the Field. New York: Teachers College Press.

Flower, Linda

    1979   Writer-Based Prose: A Cognitive Basis for Problems in Writing. College English 41:19–37.

Flower, Linda, and John R. Hayes

    1981   A Cognitive Process Theory of Writing. College Composition and Communication 32:365–387.

Fontana, Andrea

    1999   *Review of* The Art of Fieldwork, by Harry F. Wolcott. Contemporary Sociology 28(5):632–633.

Foster, George M.

    1969   Applied Anthropology. Boston: Little, Brown.

Frake, Charles O.

    1977   Plying Frames Can Be Dangerous. Quarterly Newsletter of the Institute for Comparative Human Development 1(3):1–7.

Galtung, Johan

    1990   Theory Formation in Social Research: A Plea for Pluralism. *In* Comparative Methodology. Else Øyen, ed. Pp. 96–112. Newbury Park, CA: Sage.

Geertz, Clifford

    1973   Thick Description. *In* The Interpretation of Cultures. New York: Basic Books.

    1988   Works and Lives: The Anthropologist as Author. Stanford, CA: Stanford University Press.

2000   Preface to the reissue of The Interpretation of Cultures. New York: Basic Books.

Germano, William
2001   Getting It Published: A Guide for Scholars and Anyone Else Serious about Serious Books. Chicago: University of Chicago Press.

Gills, Scott, ed.
1998   Doing Ethnographic Research: Fieldwork Settings. Thousand Oaks, CA: Sage.

Goodall, H. L (Bud)
2008   Writing Qualitative Inquiry: Self, Stories, and Academic Life. Walnut Creek, CA: Left Coast Press.

Graue, Elizabeth, and Carl Grant, eds.
2000   Review of Educational Research 69(4). [Special issue on the role of reviews in educational research.]

Gruber, Howard E.
1981   Darwin on Man. Chicago: University of Chicago Press.

Gupta, Akhil, and James Ferguson, eds.
1997   Anthropological Locations. Berkeley: University of California Press.

Hacker, Diana
2004   Rules for Writers. Boston: Bedford/St. Martin's.

Hacker, Diana, and Betty Renshaw
1979   A Practical Guide for Writers. Cambridge, MA: Winthrop Publishers.

HFW (See Wolcott, Harry F.)

Hill, Lyle Benjamin
1993   Japanese Students at an American University in Japan: An Ethnography. Unpublished Ph.D. dissertation, College of Education, University of Oregon, Eugene.

Keesing, Roger M., and Felix M. Keesing
1971   New Perspectives in Cultural Anthropology. New York: Holt, Rinehart and Winston.

Kleinman, Arthur

1995   Writing at the Margins: Discourse Between Anthropology and Medicine. Berkeley: University of California Press.

Kratz, Corrine A.

1994   On Telling/Selling a Book by Its Cover. Cultural Anthropology 9(2):179–200.

Kuper, Adam

1999   Culture: The Anthropologist's Account. Cambridge, MA: Harvard University Press.

Lamott, Anne

1994   Bird by Bird: Some Instructions on Writing and Life. New York: Pantheon Books.

Layton, Robert

1997   An Introduction to Theory in Anthropology. New York: Cambridge University Press.

LeCompte, Margaret D., Wendy L. Millroy, and Judith Preissle, eds.

1992   Handbook of Qualitative Research in Education. San Diego, CA: Academic Press.

Lewins, Ann

2007   Using Software in Qualitative Research: A Step-by-Step Guide. London: Sage.

Lewis, Oscar

1961   The Children of Sanchez: Autobiography of a Mexican Family. New York: Random House.

Lincoln, Yvonna, and Egon Guba

1985   Naturalistic Inquiry. Beverly Hills, CA: Sage.

Lomask, Milton

1987   The Biographer's Craft. New York: Harper and Row.

Malinowski, Bronislaw

1922   Argonauts of the Western Pacific: An Account of Native Enterprise and Adventure in the Archipelagoes of Melanesian New Guinea. London: Routledge and Sons.

Marcus, George E., and Michael M. Fischer

 1986   Anthropology as Cultural Critique: An Experimental Moment in the Human Sciences. Chicago: University of Chicago Press.

Markham, Annette, and Nancy K. Baym

 2009   Internet Inquiry: Conversations About Method. Thousand Oaks, CA: Sage.

Maxwell, Joseph

 2005   Qualitative Research Design: An Interactive Approach. Thousand Oaks, CA: Sage.

McCall, Michal M.

 2000   Performance Ethnography. *In* Handbook of Qualitative Research. 2nd edition. Norman K. Denzin and Yvonna Lincoln, eds. Pp. 421–433. Thousand Oaks, CA: Sage.

McDermott, Ray

 1976   Kids Make Sense: An Ethnographic Account of the Interactional Management of Success and Failure in One First-Grade Classroom. Unpublished Ph.D. dissertation, Stanford University.

Meloy, Judith M.

 1993   Writing the Qualitative Dissertation: Understanding by Doing. Hillsdale, NJ: Lawrence Erlbaum.

Merriam, Sharan B.

 1998   Qualitative Research and Case Study Applications in Education. San Francisco: Jossey-Bass.

Miles, Matthew B., and A. Michael Huberman

 1984   Qualitative Data Analysis. Beverly Hills, CA: Sage.

 1994   Qualitative Data Analysis. 2nd edition. Thousand Oaks, CA: Sage.

Mills, Geoffrey

 2007   Action Research: A Guide for the Teacher Researcher. 3rd edition. Upper Saddle River, NJ: Merrill Prentice Hall.

Murdock, George Peter

 1971   Anthropology's Mythology. Proceedings of the Royal Anthropological Institute of Great Britain and Ireland for 1971: 17–24.

Nash, Jeffrey

  1990   Working at and Working: Computer Fritters. Journal of
  Contemporary Ethnography 19(2):207–225.

Noblit, George W., and R. Dwight Hare

  1988   Meta-Ethnography: Synthesizing Qualitative Studies. Sage
  Qualitative Research Methods Series, Volume 11. Newbury Park,
  CA: Sage.

Pan, M. Ling

  2008   Preparing Literature Reviews: Qualitative and Quantitative
  Approaches. 3rd edition. Glendale, CA: Pyrczak Publishers

Peacock, James L.

  1986   The Anthropological Lens: Harsh Light, Soft Focus. New York:
  Cambridge University Press.

Peshkin, Alan

  1985   From Title to Title: The Evolution of Perspective in Naturalistic
  Inquiry. Anthropology and Education Quarterly 16(3):214–224.

Powell, Walter W.

  1985   Getting Into Print: The Decision-Making Process in Scholarly
  Publishing. Chicago: University of Chicago Press.

Richardson, Laurel

  1990   Writing Strategies: Reaching Diverse Audiences. Sage
  Qualitative Research Methods Series, Volume 21. Newbury Park,
  CA: Sage.

  2000   Writing: A Method of Inquiry. *In* Handbook of Qualitative
  Research. 2nd edition. Norman K. Denzin and Yvonna S. Lincoln,
  eds. Pp. 923–948. Thousand Oaks, CA: Sage.

Rohner, Ronald P.

  1975   They Love Me, They Love Me Not: A Worldwide Study of the
  Effects of Parental Acceptance and Rejection. New Haven, CT: HRAF
  Press.

Rohner, Ronald P., and Evelyn C. Rohner

  1970   The Kwakiutl Indians of British Columbia. New York: Holt,
  Rinehart and Winston. [Reissued 1986 by Waveland Press.]

Rubin, Herbert J.

  2005   Qualitative Interviewing: The Art of Hearing Data. Thousand
  Oaks, CA: Sage.

Rudestam, Kjell Erik, and Rae Newton

2007   Surviving Your Dissertation: A Comprehensive Guide to Content and Process. 3rd edition. Thousand Oaks, CA: Sage.

Ryan, Gery, and H. Russell Bernard

2000   Data Management and Analysis Methods. *In* Handbook of Qualitative Research. 2nd edition. Norman K. Denzin and Yvonna S. Lincoln, eds. Pp. 769–802. Thousand Oaks, CA: Sage.

Saldaña, Johnny, ed.

2005   Ethnodrama: An Anthology of Reality Theatre. Lanham, MD: Rowman and Littlefield.

Salinas Pedraza, Jesús, in collaboration with H. Russell Bernard

1978   Rc Hnychnyu: The Otomí. Volume 1: Geography and Fauna. Albuquerque: University of New Mexico Press.

Sanjek, Roger

1999   Into the Future: Is Anyone Listening? Paper presented at the American Anthropological Association meetings, Chicago.

Seidman, Irving

1991   Interviewing as Qualitative Research: A Guide for Researchers in the Social Sciences. New York: Teachers College Press.

Shostak, Marjorie

1981   Nisa: The Life and Words of a !Kung Woman. New York. Random House.

Simmons, Leo W., ed.

1942   Sun Chief: The Autobiography of a Hopi Indian. New Haven, CT: Yale University Press.

Smedley, Christine S., Mitchell Allen, and Associates

1993   Getting Your Book Published. Newbury Park, CA: Sage.

Spector, Janet D.

1993   What This Awl Means: Feminist Archaeology at a Wahpeton Dakota Village. St. Paul: Minnesota Historical Society Press.

Stake, Robert E.

1995   The Art of Case Study Research. Thousand Oaks, CA: Sage.

Stein, Gertrude

1937   Everybody's Autobiography. New York: Random House.

Stevens, Michael E., and Steven B. Burg

1997   Editing Historical Documents: A Handbook of Practice. Walnut Creek, CA: AltaMira Press.

Strunk, William, Jr.

1918   The Elements of Style. [Paper written for students at Cornell University and published privately.]

Strunk, William, Jr., and Edward A. Tenney

1934   The Elements of Style. New York: Harcourt, Brace and Co.

Strunk, William, Jr., and E. B. White

1972   Elements of Style. 2nd edition. New York. Macmillan.

Sturges, Keith

Forthcoming. "Lessons Learned: Identity Production of the Modern Program Evaluator." Working Ph.D. dissertation title, Cultural Studies in Education Program, University of Texas.

Taylor, Steven J., and Robert Bogdan

1984   Introduction to Qualitative Research Methods: The Search for Meanings. 2nd edition. New York: John Wiley and Sons.

Thyer, Bruce A.

1994   Successful Publishing in Scholarly Journals. Thousand Oaks, CA: Sage.

Tufte, Edward R.

1983   The Visual Display of Quantitative Information. Cheshire, CT: Graphics Press.

1990   Envisioning Information. Cheshire, CT: Graphics Press.

Van Maanen, John

1988   Tales of the Field: On Writing Ethnography. Chicago: University of Chicago Press.

Wallgren, Anders, Britt Wallgren, Rolf Persson, Ulf Jorner, and Jan-Aage Haaland.

1996   Graphing Statistics & Data. Thousand Oaks, CA: Sage.

Wax, Murray L., Rosalie H. Wax, and Robert V. Dumont, Jr.

1964   Formal Education in an American Indian Community. Supplement to Social Problems 11(4). [Reissued with changes 1989 by Waveland Press.]

Wax, Rosalie H.

1971   Doing Fieldwork: Warnings and Advice. Chicago: University of Chicago Press.

Weitzman, Eben A.

2000   Software and Qualitative Research. *In* Handbook of Qualitative Research. 2nd edition. Norman K. Denzin and Yvonna S. Lincoln, eds. Pp. 803–820. Thousand Oaks, CA: Sage.

Whyte, William F.

1943   Street Corner Society: The Social Structure of an Italian Slum. Chicago: University of Chicago Press.

1955   Street Corner Society: The Social Structure of an Italian Slum. 2nd edition, enlarged. Chicago: University of Chicago Press.

Wolcott, Harry F.

1964   A Kwakiutl Village and Its School. Unpublished Ph.D. dissertation, Stanford University.

1967   A Kwakiutl Village and School. New York: Holt, Rinehart and Winston. [Reissued 1989 by Waveland Press with a new Afterword.]

1973   The Man in the Principal's Office: An Ethnography. New York: Holt, Rinehart and Winston. [Reissued 1984 by Waveland Press with a new Introduction.]

1974a   The African Beer Gardens of Bulawayo: Integrated Drinking in a Segregated Society. New Brunswick, NJ: Rutgers Center of Alcohol Studies. Monograph Number 10.

1974b   The Elementary School Principal: Notes from a Field Study. *In* Education and Cultural Process: Toward an Anthropology of Education. George D. Spindler, ed. Pp. 176–204. New York: Holt, Rinehart and Winston. [Reissued 1987, 1997 by Waveland Press.]

1975   Introduction. [Special issue on the Ethnography of Schooling.] Human Organization 34(2):109–110.

1977 Teachers Versus Technocrats: An Educational Innovation in Anthropological Perspective. Eugene: Center for Educational Policy and Management, University of Oregon.

1982 Differing Styles of On-Site Research, Or, "If It Isn't Ethnography, What Is It?" Review Journal of Philosophy and Social Science 7(1,2):154–169.

1983a Adequate Schools and Inadequate Education: The Life History of a Sneaky Kid. Anthropology and Education Quarterly 14(1): 3–32. [Reprinted in Transforming Qualitative Data: Description, Analysis, and Interpretation. Thousand Oaks, CA: Sage, 1994.]

1983b A Malay Village That Progress Chose: Sungai Lui and the Institute of Cultural Affairs. Human Organization 42(1): 72–81. [Reprinted in Transforming Qualitative Data: Description, Analysis, and Interpretation. Thousand Oaks, CA: Sage, 1994.]

1987 On Ethnographic Intent. *In* Interpretive Ethnography of Education. George and Louise Spindler, eds. Pp. 37–57. Hillsdale, NJ: Lawrence Erlbaum.

1988 Problem Finding in Qualitative Research. *In* School and Society: Learning Content through Culture. Henry Trueba and C. Delgado-Gaitan, eds. Pp. 11–35. New York: Praeger.

1990a Making a Study "More Ethnographic." Journal of Contemporary Ethnography 19(1): 79–111. [Special Issue. The Presentation of Ethnographic Research.]

1990b Writing Up Qualitative Research. Sage Qualitative Research Methods Series, Volume 20. Newbury Park, CA: Sage.

1991 The Acquisition of Culture: Notes on a Working Paper. *In* Diversity and Design: Studying Culture and the Individual. Mary Jo McGee-Brown, ed. Pp. 22–46. Athens: Proceedings of the Fourth Annual Conference of the Qualitative Research Group, University of Georgia. [Reprinted in Transforming Qualitative Data: Description, Analysis, and Interpretation. Thousand Oaks, CA: Sage, 1994.]

1994 Transforming Qualitative Data: Description, Analysis, and Interpretation. Thousand Oaks, CA: Sage.

1995 The Art of Fieldwork. Walnut Creek, CA: AltaMira Press.

1999a Ethnography: A Way of Seeing. Walnut Creek, CA: AltaMira Press.

1999b Electronic review of Learning in Likely Places: Varieties of Apprenticeship in Japan. John Singleton, ed. New York: Cambridge

University Press 1998. 376 pages. Indexed in Anthropology and Education Quarterly, Vol 32 (2): 261. Retrieved 31 March 08. http://dev.aaanet.org/sections/cae/aeq/br/singleton2.htm

2001   Writing Up Qualitative Research. 2nd edition. Thousand Oaks, CA: Sage.

2002   Sneaky Kid and Its Aftermath: Ethics and Intimacy in Fieldwork. Walnut Creek, CA: AltaMira Press.

2008a   Ethnography: A Way of Seeing. 2nd edition. Lanham, MD: Rowman and Littlefield.

2008b   *Electronic Review of* Telling about Society by Howard S. Becker. Teachers College Record. Retrieved 3/20/08: http://www.tcrecord .org/Content.asp?ContentID=14871.

Woods, Peter

1985   New Songs Played Skillfully: Creativity and Technique in Writing Up Qualitative Research. *In* Issues in Educational Research: Qualitative Methods. Robert G. Burgess, ed. Pp. 86–106. Philadelphia, PA: Falmer Press.

1999   Successful Writing for Qualitative Researchers. London: Routledge.

Yin, Robert K.

1994   Case Study Research: Design and Methods. Thousand Oaks, CA: Sage.

2008   Case Study Research: Design and Methods. 4th ed. Thousand Oaks, CA: Sage.

Zinsser, William K

1976   On Writing Well: An Informal Guide to Writing Nonfiction. New York: Harper & Row.

1998   On Writing Well: The Classic Guide to Writing Nonfiction. 6th edition. New York: HarperCollins.

Name Index

Subject Index

DATE DUE